T0342283

ORGANIC SYNTHESES

ORGANIC SYNTHESES

AN ANNUAL PUBLICATION OF SATISFACTORY METHODS FOR THE PREPARATION OF ORGANIC CHEMICALS

VOLUME 80

2003

A JOHN WILEY & SONS, INC., PUBLICATION

Published by John Wiley & Sons, Inc., Hoboken, New Jersey.

Published simultaneously in Canada.

Library of Congress Catalog Card Number: 21-17747
ISBN 978-0-471-44531-9

ORGANIC SYNTHESES

VOLUME	EDITOR-IN-CHIEF	PAGES
I*	†Roger Adams	84
II*	†James Bryant Conant	100
III*	†Hans Thacher Clarke	105
IV*	†Oliver Kamm	89
V*	†Carl Shipp Marvel	110
VI*	†Henry Gilman	120
VII*	†Frank C. Whitmore	105
VIII*	†Roger Adams	139
IX*	†James Bryant Conant	108
Collective Vol. I	A revised edition of Annual Volumes I–IX †Henry Gilman, *Editor-in-Chief* 2nd Edition revised by †A. H. Blatt	580
X*	†Hans Thacher Clarke	119
XI*	†Carl Shipp Marvel	106
XII*	†Frank C. Whitmore	96
XIII*	†Wallace H. Carothers	119
XIV*	†William W. Hartman	100
XV*	†Carl R. Noller	104
XVI*	†John R. Johnson	104
XVII*	†L. F. Fieser	112
XVIII*	†Reynold C. Fuson	103
XIX*	†John R. Johnson	105
Collective Vol. II	A revised edition of Annual Volumes X–XIX †A. H. Blatt, *Editor-in-Chief*	654
20*	†Charles F. H. Allen	113
21*	†Nathan L. Drake	120
22*	†Lee Irvin Smith	114
23*	†Lee Irvin Smith	124
24*	†Nathan L. Drake	119
25*	†Werner E. Bachmann	120
26*	†Homer Adkins	124
27*	†R. L. Shriner	121
28*	†H. R. Snyder	121
29*	†Cliff S. Hamilton	119
Collective Vol. III	A revised edition of Annual Volumes 20–29 †E. C. Horning, *Editor-in-Chief*	890
30*	†Arthur C. Cope	115
31*	†R. S. Schreiber	122

**Out of print.*
† Deceased.

VOLUME	EDITOR-IN-CHIEF	PAGES
32*	†Richard T. Arnold	119
33*	†Charles C. Price	115
34*	†William S. Johnson	121
35*	†T. L. Cairns	122
36*	N. J. Leonard	120
37*	James Cason	109
38*	†John C. Sheehan	120
39*	†Max Tishler	114
Collective Vol. IV	A revised edition of Annual Volumes 30–39 †Norman Rabjohn, *Editor-in-Chief*	1036
40*	†Melvin S. Newman	114
41*	John D. Roberts	118
42*	Virgil Boekelheide	118
43*	B. C. McKusick	124
44*	†William E. Parham	131
45*	†William G. Dauben	118
46*	E. J. Corey	146
47*	†William D. Emmons	140
48*	†Peter Yates	164
49*	Kenneth B. Wiberg	124
Collective Vol. V	A revised edition of Annual Volumes 40–49 Henry E. Baumgarten, *Editor-in-Chief*	1234
Cumulative Indices to Collective Volumes, I, II, III, IV, V	†Ralph L. and †Rachel H. Shriner, *Editors*	
50*	Ronald Breslow	136
51*	†Richard E. Benson	209
52*	Herbert O. House	192
53*	Arnold Brossi	193
54*	Robert E. Ireland	155
55*	Satoru Masamune	150
56*	†George H. Büchi	144
57*	Carl R. Johnson	135
58*	†William A. Sheppard	216
59*	Robert M. Coates	267
Collective Vol. VI	A revised edition of Annual Volumes 50–59 Wayland E. Noland, *Editor-in-Chief*	1208
60*	Orville L. Chapman	140
61*	†Robert V. Stevens	165

****Out of print.***
†Deceased.

VOLUME	EDITOR-IN-CHIEF	PAGES
62*	MARTIN F. SEMMELHACK	269
63*	GABRIEL SAUCY	291
64	ANDREW S. KENDE	308
Collective Vol. VII	A revised edition of Annual Volumes 60–64 JEREMIAH P. FREEMAN, *Editor-in-Chief*	602
65	EDWIN VEDEJS	278
66*	CLAYTON H. HEATHCOCK	265
67	BRUCE E. SMART	289
68	JAMES D. WHITE	318
69	LEO A. PAQUETTE	328
Collective Vol. VIII	A revised edition of Annual Volumes 65–69 JEREMIAH P. FREEMAN, *Editor-in-Chief*	696
70	ALBERT I. MEYERS	305
71	LARRY E. OVERMAN	285
72	DAVID L. COFFEN	333
73	ROBERT K. BOECKMAN, JR.	352
74	ICHIRO SHINKAI	341
Collective Vol. IX	A revised edition of Annual Volumes 70–74 JEREMIAH P. FREEMAN, *Editor-in-Chief*	840
75	AMOS B. SMITH III	257
76	STEPHEN F. MARTIN	340
77	DAVID J. HART	312
78	WILLIAM R. ROUSH	326
79	LOUIS S. HEGEDUS	328
80	STEVEN WOLFF	259

Collective Volumes, Collective Indices to Collective Volumes I–VIII, Volumes 75–79, and Reaction Guide are available from John Wiley & Sons, Inc.

**Out of print.*
†*Deceased.*

NOTICE

With Volume 62, the Editors of *Organic Syntheses* began a new presentation and distribution policy to shorten the time between submission and appearance of an accepted procedure. The soft cover edition of this volume is produced by a rapid and inexpensive process, and is sent at no charge to members of the Organic Division of the American Chemical Society, The Perkin Division of the Royal Society of Chemistry, and The Society of Synthetic Organic Chemistry, Japan. The soft cover edition is intended as the personal copy of the owner and is not for library use. The hard cover edition is published by John Wiley and Sons, Inc., in the traditional format, and it differs in content primarily by the inclusion of an index. The hard cover edition is intended primarily for library collections and is available for purchase through the publisher. Incorporation of graphical abstracts into the Table of Contents began with Volume 77. Annual Volumes 70–74 have been incorporated into the new five-year version of the collective volumes of *Organic Syntheses* that appeared as *Collective Volume Nine* in the traditional hard cover format. It is available for purchase from the publishers. The Editors hope that the new *Collective Volume* series, appearing twice as frequently as the previous decennial volumes, will provide a permanent and timely edition of the procedures for personal and institutional libraries. The Editors welcome comments and suggestions from users concerning the new editions.

Organic Syntheses, Inc., joined the age of electronic publication in 2001 with the release of its free web site, and now in 2003 with the completion of a commercially available electronic database (www.mrw.interscience.wiley.com/osdb). John Wiley & Sons, Inc., and Accelrys, Inc., partnered with Organic Syntheses, Inc. to develop the new database that is available for license with internet solutions from John Wiley & Sons, Inc., and intranet solutions offered in a few months by Accelrys, Inc. Organic Syntheses, Inc., fully funded the creation of the free website at www.orgsyn.org in a partnership with CambridgeSoft Corporation, and Data-Trace Publishing Company. The success of this site in its first full year of operation was overwhelming, with an average of nearly 48,000 site hits/day and more than 27,000 page views/day. Also noteworthy is the imminent release of ChemDraw Active X/Plugin Net 8.0 that enables users with Macintosh Operating System to access the site. However since some Macintosh browser limitations remain, Mac users will need to carefully review software compatibility at the site prior to use.

Both the commercial database and the free website contain all nine collective volumes, annual volumes 1–80, and indices of *Organic Syntheses.* They are PC and Mac compatible, and are searchable using a variety of techniques. Chemists can draw structural queries and combine structural or reaction transformation queries with full-text and bibliographic search terms, such as chemical name, reagents, molecular formula, apparatus, or even hazard warnings or phrases. The preparations are categorized into reaction types, allowing search by category. The contents of individual or collective volumes can be browsed by lists of titles, submitters' names, and volume and page references, with or without reaction equations.

The commercial database at www.mrw.interscience.wiley.com/osdb also enables the user to choose his/her preferred chemical drawing package, or to utilize several freely available plug-ins for entering queries. The user is also able to cut and paste existing structures and reactions directly into the structure search query or their preferred chemistry editor, streamlining workflow. Additionally, this database contains links to the full text of primary literature references via CrossRef, ChemPort, Medline, and ISI Web of Science. Links to local holdings for institutions using open url technology can also be enabled. The database user can limit his/her search to, or order the search results by, such factors as reaction type, percentage yield, temperature, and publication date, and can create a customized table of reactions for comparison. Connections to other Wiley references are currently made via text search, with cross-product structure and reaction searching to be added in the coming year. Inclusion of the reactions from the next annual volume into the database, and the posting of the contents on the website, will occur as new material becomes available.

NOMENCLATURE

Both common and systematic names of compounds are used throughout this volume, depending on which the Editor-in-Chief felt was more appropriate. The *Chemical Abstracts* indexing name for each title compound, if it differs from the title name, is given as a subtitle. Systematic *Chemical Abstracts* nomenclature, used in the Collective Indexes for the title compound and a selection of other compounds mentioned in the procedure, is provided in an appendix at the end of each preparation. Registry numbers, which are useful in computer searching and identification, are also provided in these appendixes. Whenever two names are concurrently in use and one name is the correct *Chemical Abstracts* name, that name is preferred.

SUBMISSION OF PREPARATIONS

Organic Syntheses welcomes and encourages submission of experimental procedures which lead to compounds of wide interest or which illustrate important new developments in methodology. The Editorial Board will consider proposals in outline format as shown below, and will request full experimental details for those proposals which are of sufficient interest. Tear-out copies of this form may be found at the back of this volume. Submissions which are longer than three steps from commercial sources or from existing *Organic Syntheses* procedures will be accepted only in unusual circumstances.

Proposals and Procedures may be submitted in electronic form compatible with Macintosh.

Organic Syntheses Proposal Format

1) Authors
2) Title
3) Literature reference or enclose preprint if available
4) Proposed sequence
5) Best current alternative(s)
6) a. Proposed scale, final product:
   b. Overall yield:
   c. Method of isolation and purification:

d. Purity of product (%):
e. How determined?
7) Any unusual apparatus or experimental technique?
8) Any hazards?
9) Source of starting material?
10) Utility of method or usefulness of product

Submit to: Dr. Jeremiah P. Freeman, Secretary
Department of Chemistry
University of Notre Dame
Notre Dame, IN 46556

Proposals will be evaluated in outline form, again after submission of full experimental details and discussion, and, finally by checking experimental procedures. A form that details the preparation of a complete procedure (Notice to Submitters) may be obtained from the Secretary.

Additions, corrections, and improvements to the preparations previously published are welcomed; these should be directed to the Secretary. However, checking of such improvements will only be undertaken when new methodology is involved. Substantially improved procedures have been included in the Collective Volumes in place of a previously published procedure.

ACKNOWLEDGMENT

Organic Syntheses wishes to acknowledge the contributions of Discovery Partners Intl., Hoffmann-La Roche, Inc., Merck & Co., and Pfizer, Inc. to the success of this enterprise through their support, in the form of time and expenses, of members of the Boards of Directors and Editors.

HANDLING HAZARDOUS CHEMICALS
A Brief Introduction

General Reference: *Prudent Practices in the Laboratory*; National Academy Press; Washington, DC, 1995.

Physical Hazards

Fire. Avoid open flames by use of electric heaters. Limit the quantity of flammable liquids stored in the laboratory. Motors should be of the nonsparking induction type.

Explosion. Use shielding when working with explosive classes such as acetylides, azides, ozonides, and peroxides. Peroxidizable substances such as ethers and alkenes, when stored for a long time, should be tested for peroxides before use. Only sparkless "flammable storage" refrigerators should be used in laboratories.

Electric Shock. Use 3-prong grounded electrical equipment if possible.

Chemical Hazards

Because all chemicals are toxic under some conditions, and relatively few have been thoroughly tested, it is good strategy to minimize exposure to all chemicals. In practice this means having a good, properly installed hood; checking its performance periodically; using it properly; carrying out all operations in the hood; protecting the eyes; and, since many chemicals can penetrate the skin, avoiding skin contact by use of gloves and other protective clothing at all times.

a. Acute Effects. These effects occur soon after exposure. The effects include burn, inflammation, allergic responses, damage to the eyes, lungs, or nervous system (e.g., dizziness), and unconsciousness or death (as from overexposure to HCN). The effect and its cause are usually obvious and so are the methods to prevent it. They generally arise from inhalation or skin contact, so should not be a problem if one follows the admonition "work in a hood and keep chemicals off your hands." Ingestion is a rare route, being generally the result of eating in the laboratory or not washing hands before eating.

b. Chronic Effects. These effects occur after a long period of exposure or after a long latency period and may show up in any of numerous organs. Of the chronic effects of chemicals, cancer has received the most attention lately. Several dozen chemicals have been demonstrated to be carcinogenic in man and hundreds to be carcinogenic to animals. Although there is no simple correlation between carcinogenicity in animals and in man, there is little doubt that a significant proportion of the chemicals used in laboratories have some potential for carcinogenicity in man. For this and other reasons, chemists should employ good practices at all times.

The key to safe handling of chemicals is a good, properly installed hood, and the referenced book devotes many pages to hoods and ventilation. It recommends that in a laboratory where people spend much of their time working with chemicals there should be a hood for each two people, and each should have at least 2.5 linear feet (0.75 meter) of working space at it. Hoods are more than just devices to keep undesirable vapors from the laboratory atmosphere. When closed they provide a protective barrier between chemists and chemical operations, and they are a good containment device for spills. Portable shields can be a useful supplement to hoods, or can be an alternative for hazards of limited severity, e.g., for small-scale operations with oxidizing or explosive chemicals.

Specialized equipment can minimize exposure to the hazards of laboratory operations. Impact resistant safety glasses are basic equipment and should be worn at all times. They may be supplemented by face shields or goggles for particular operations, such as pouring corrosive liquids. Because skin contact with chemicals can lead to skin irritation or sensitization or, through absorption, to effects on internal organs, protective gloves should be worn at all times.

Laboratories should have fire extinguishers and safety showers. Respirators should be available for emergencies. Emergency equipment should be kept in a central location and must be inspected periodically.

MSDS (Materials Safety Data Sheets) are available from the suppliers of commercially available reagents, solvents, and other chemical materials; anyone performing an experiment should check these data sheets before initiating an experiment to learn of any specific hazards associated with the chemicals being used in that experiment.

DISPOSAL OF CHEMICAL WASTE

General Reference: *Prudent Practices in the Laboratory*, National Academy Press, Washington, D.C. 1995

Effluents from synthetic organic chemistry fall into the following categories:

1. **Gases**

   1a. Gaseous materials either used or generated in an organic reaction.
   1b. Solvent vapors generated in reactions swept with an inert gas and during solvent stripping operations.
   1c. Vapors from volatile reagents, intermediates and products.

2. **Liquids**

   2a. Waste solvents and solvent solutions of organic solids (see item 3b).
   2b. Aqueous layers from reaction work-up containing volatile organic solvents.
   2c. Aqueous waste containing non-volatile organic materials.
   2d. Aqueous waste containing inorganic materials.

3. **Solids**

   3a. Metal salts and other inorganic materials.
   3b. Organic residues (tars) and other unwanted organic materials.
   3c. Used silica gel, charcoal, filter aids, spent catalysts and the like.

The operation of industrial scale synthetic organic chemistry in an environmentally acceptable manner* requires that all these effluent categories be dealt with properly. In small scale operations in a research or academic set-

*An environmentally acceptable manner may be defined as being both in compliance with all relevant state and federal environmental regulations *and* in accord with the common sense and good judgement of an environmentally aware professional.

ting, provision should be made for dealing with the more environmentally offensive categories.

1a. Gaseous materials that are toxic or noxious, e.g., halogens, hydrogen halides, hydrogen sulfide, ammonia, hydrogen cyanide, phosphine, nitrogen oxides, metal carbonyls, and the like.
1c. Vapors from noxious volatile organic compounds, e.g., mercaptans, sulfides, volatile amines, acrolein, acrylates, and the like.
2a. All waste solvents and solvent solutions of organic waste.
2c. Aqueous waste containing dissolved organic material known to be toxic.
2d. Aqueous waste containing dissolved inorganic material known to be toxic, particularly compounds of metals such as arsenic, beryllium, chromium, lead, manganese, mercury, nickel, and selenium.
3. All types of solid chemical waste.

Statutory procedures for waste and effluent management take precedence over any other methods. However, for operations in which compliance with statutory regulations is exempt or inapplicable because of scale or other circumstances, the following suggestions may be helpful.

Gases

Noxious gases and vapors from volatile compounds are best dealt with at the point of generation by "scrubbing" the effluent gas. The gas being swept from a reaction set-up is led through tubing to a (large!) trap to prevent suck-back and on into a sintered glass gas dispersion tube immersed in the scrubbing fluid. A bleach container can be conveniently used as a vessel for the scrubbing fluid. The nature of the effluent determines which of four common fluids should be used: dilute sulfuric acid, dilute alkali or sodium carbonate solution, laundry bleach when an oxidizing scrubber is needed, and sodium thiosulfate solution or diluted alkaline sodium borohydride when a reducing scrubber is needed. Ice should be added if an exotherm is anticipated.

Larger scale operations may require the use of a pH meter or starch/iodide test paper to ensure that the scrubbing capacity is not being exceeded.

When the operation is complete, the contents of the scrubber should be handled as aqueous waste, as outlined in the "Liquids" section that follows. In many instances, this will require neutralization, followed by concentration to a minimum volume, or concentration to dryness before disposal as concentrated liquid or solid chemical waste.

Liquids

Every laboratory should be equipped with a waste solvent container in which *all* waste organic solvents and solutions are collected. The contents of these containers should be periodically transferred to properly labeled waste solvent drums and arrangements made for contracted disposal in a regulated and licensed incineration facility.**

Aqueous waste containing dissolved toxic organic material should be decomposed *in situ*, when feasible, by adding acid, base, oxidant, or reductant. Otherwise, the material should be concentrated to a minimum volume and added to the contents of a waste solvent drum.

Aqueous waste containing dissolved toxic inorganic material should be evaporated to dryness and the residue handled as a solid chemical waste.

Solids

Soluble organic solid waste can usually be transferred into a waste solvent drum, provided near-term incineration of the contents is assured.

Inorganic solid wastes, particularly those containing toxic metals and toxic metal compounds, used Raney nickel, manganese dioxide, etc. should be placed in glass bottles or lined fiber drums, sealed, properly labeled, and arrangements made for disposal in a secure landfill.** Used mercury is particularly pernicious and small amounts should first be amalgamated with zinc or combined with excess sulfur to solidify the material.

Other types of solid laboratory waste including used silica gel and charcoal should also be packed, labeled, and sent for disposal in a secure landfill.

Special Note

Since local ordinances may vary widely from one locale to another, one should always check with appropriate authorities. Also, professional disposal services differ in their requirements for segregating and packaging waste.

**If arrangements for incineration of waste solvent and disposal of solid chemical waste by licensed contract disposal services are not in place, a list of providers of such services should be available from a state or local office of environmental protection.

PREFACE

The 80th annual volume of *Organic Syntheses* once again demonstrates the creativity, ingenuity, and vitality of the practitioners of this art. This collection convincingly refutes whatever suspicion persists that this discipline, while clearly mature, has become passé. The current volume continues the tradition of providing significant, carefully checked and edited, and interesting procedures, which should prove worthwhile to the many synthetic chemists working in increasingly diverse areas. Following precedent, there is no specific or central theme to this volume, but the procedures can be organized into four main categories: (1) asymmetric syntheses, chiral auxiliaries and chiral ligands; (2) reactions that are promoted by metals or proceed via organometallic intermediates; (3) cycloaddition reactions; (4) valuable synthetic transformations and building blocks.

The compilation begins with the straightforward preparation of **1,2:4,5-DI-O-ISOPROPYLIDENE-D-erythro-2,3-HEXODIULO-2,6-PYRANOSE** and its application in the enantioselective syntheses of **(R,R)-trans-β-METHYLSTYRENE OXIDE** and **(R,R)-1-PHENYLCYCLOHEXENE OXIDE** via the intermediacy of a chiral dioxirane. A second set of procedures for the asymmetric synthesis of 4-tert-butoxycarbonyl α-amino acids, specifically **(R)-(N-tert-BUTOXYCARBONYL)ALLYLGLYCINE** prepared by the alkylation of **(5S,6R)-4-tert-BUTOXYCARBONYL-5,6-DIPHENYLMORPHOLIN-2-ONE** with allyl iodide follows. The preparation of the phase transfer catalyst **O-ALLYL-N-(9-ANTHRACENYLMETHYL)CINCHONIDINIUM BROMIDE** and its utilization in the enantioselective Michael addition of glycine benzophenone imine to methyl acrylate to afford **(4S)-2-(BENZHYDRYLIDENAMINO)PENTANOIC ACID, 1-tert BUTYL ESTER-5-METHYL ESTER** is then reported. Using this catalyst, the imine can also be alkylated with an assortment of electrophiles, including aliphatic, allylic and benzylic halides as well as cyclic and acylic α,β-unsaturated esters and ketones. The series continues with the convenient, four-step preparation of **(R,R)-4,6-DIBENZOFURANDIYL-2,2′-BIS(4-PHENYLOXAZOLINE)**, an extremely useful ligand for a variety of asymmetric transformations including Diels-Alder reactions and nitrone cycloadditions. The preparation of **(4S)-4-(1-METHYLETHYL)-5,5-DIPHENYL-2-OXAZOLIDINONE**, a chiral auxiliary derived from L-valine, which may possess several advantages over similar, but somewhat dated oxazolidinones, is detailed. Finally, the synthesis of the chiral butenolide, **(5S)-(d-MENTHYLOXY)-2(5H)-FURANONE**, whose conjugated double bond readily participates in Diels-Alder reactions, 1,4-conjugate additions, and [2+2] photochemical additions, among others, concludes this set of procedures.

The next series of seven procedures provides examples of metal- and organometallic-promoted processes and illustrate how valuable, indeed commonplace, they have become in synthetic practice. The preparation of **METHYL INDOLE-4-CARBOXYLATE** by the palladium-catalyzed reductive N-heteroannulation of 2-nitrostyrenes is an efficient alternative to other methods previously reported in these volumes and permits access to indoles substituted in the 2- and/or 3-position. A procedure for the synthesis of **N-Boc-3-PYRROLINE** illustrates the ring-closing metathesis (RCM) reaction—a remarkable process whose impact on organic synthesis requires no comment—of N-Boc-diallylamine and the very practical removal of ruthenium-containing impurities with tris(hydroxymethyl)phosphine. An example of one of the versions of the catalytic intramolecular Pauson-Khand reaction using a complex formed from dicobalt octacarbonyl and 2-methyl-3-butyn-2-ol to afford **2,3,3α,4-TETRAHYDRO-2-[(4-METHYLBENZENE)SULFONYL]CYCLOPENTA[C]PYRROL-5(1H)-ONE** is then described. The preparation of **8-[2-(TRIETHOXYSILYL)ETHYL]-1-TETRALONE**, an atypical example of the ruthenium-catalyzed alkylation of aromatic ketones at the ortho position with olefins, follows. This reaction is quite general and high yielding, and a wide variety of functional groups on the aromatic ring is tolerated. The synthesis of **BICYCLO[3.1.0]HEXAN-1-OL** and **trans-2-BENZYL-1-METHYLCYCLOPROPAN-1-OL** serves to illustrate an improved modification of both the intra- and intermolecular Kulinkovich cyclopropanation reaction of carboxylic esters with olefins. The preparation of **(Z)-1,2-DIDEUTERIO-1-(TRIMETHYLSILYL)-1-HEXENE** demonstrates one of the many fates of an alkyne titanium alkoxide complex upon reaction with electrophiles, in the present case deuterium oxide. Finally, the copper(I) iodide-catalyzed addition of hydriodic acid to propiolic acid to afford stereospecifically **(E)-3-IODOPROP-2-ENOIC ACID**, a valuable three-carbon building block, is described.

A set of four cycloaddition reactions follows. The preparation of **METHYL 5-METHYLPYRIDINE-2-CARBOXYLATE** from the regiocontrolled, Lewis acid-promoted [4+2] cycloaddition of an oximinosulfonate derived from Meldrum's acid and isoprene exemplifies a novel procedure to obtain pyridines. Most importantly, the regiochemistry of this cycloaddition is opposite to that observed with standard imino dienophiles, thus permitting the synthesis of pyridines with a broader range of substitution patterns. The thermolysis of **6,6-DIMETHYL-1-METHYLENE-4,8-DIOXASPIRO-[2.5]OCTANE**, synthesized in three steps from 1,3-dichloroacetone and neopentyl glycol, in the presence of cyclopentenone yields **cis-5-(5,5-DIMETHYL-1,3-DIOXAN-2-YLIDENE)HEXAHYDRO-1(2H)-PENTALEN-2-ONE** via [3+2] cycloaddition of a trimethylenemethane intermediate. Many other unsaturated substrates successfully undergo cycloaddition with this reactive species. The generation and [2+2] cycloaddition of thio-substituted ketenes is demonstrated by the preparation of **trans-1-(4-METHOXYPHENYL)-4-PHENYL-3-(PHENYLTHIO)-AZETIDIN-2-ONE**. These ketenes react facilely with unactivated olefins and the arylthio group of the cycloadduct can conveniently be removed by reduction, thus providing an attractive, alternative process to the use of dichloroketene. Finally a novel method to generate difluorocarbene by fluoride-induced fragmentation of **TRIMETHYLSILYL 2-FLUOROSULFONYL-2,2-DIFLUOROACETATE** in the presence of n-butyl acrylate to yield **n-BUTYL 2,2-DIFLUOROCYCLOPROPANECARBOXYLATE** is described.

This collection ends with a rather varied group of synthetic methods. Among the several noteworthy transformations reported is the preparation of α-alkoxy ethers, specifically **endo-1-BORNYLOXYETHYL ACETATE** by the low temperature reduction of esters with diisobutylaluminum hydride, followed by trapping of the intermediate hemiacetal with acetic anhydride using 4-dimethylaminopyridine as catalyst. These α-alkoxy ethers easily generate oxacarbenium ions, which can subsequently be trapped by many types of nucleophiles. The selective oxidation of mercaptans and sulfides with hydrogen peroxide in trifluoroethanol, namely thiophenol to yield **DIPHENYL DISULFIDE** and thioanisole to give **PHENYL METHYL SULFOXIDE** is then detailed. The conversion of 3-phenyl-2-propyn-1-ol to **1-HYDROXY-3-PHENYL-2-PROPANONE** mediated by 1-propanethiol is a convenient method to access these functional building blocks without resorting to reductive, oxidative or metal-promoted processes. A mildly basic catalyst, diethyltrimethylsilylamine, efficiently catalyzes 1,4-addition of aldehydes to methyl or ethyl vinyl ketones, thus furnishing 1,5-dicarbonyl compounds, which are generally very useful for the preparation of substituted cyclohexenones; in this manner, **(3R)-3,7-DIMETHYL-2-(3-OXOBUTYL)-6-OCTENAL** is prepared. Treatment of dimethylformamide with chloroacetyl chloride in the presence of phosphorus oxychloride followed by formation of the hexafluorophosphate salt yields **2-CHLORO-1,3-BIS(DIMETHYLAMINO)TRIMETHINIUM HEXAFLUOROPHOSPHATE**, a practical reagent for the synthesis of pyridines and pyridones. A novel method for the conversion of primary amines into N-monoalkylhydroxylamines, in this instance, **N-HYDROXY-(S)-1-PHENYLETHYLAMINE OXALATE**, via a three-step protocol involving cyanomethylation with bromoacetonitrile, regioselective nitrone formation by oxidation with m-chloroperbenzoic acid and finally hydroxylaminolysis is detailed. The preparation of **TETRABENZYL PYROPHOSPHATE** from dibenzyl phosphate affords a frequently used reagent for the introduction of phosphate groups into organic molecules having biological importance. Finally the volume concludes with a pair of procedures that utilize rather classical techniques, namely sodium dithionite reduction of phenanthrenequinone and subsequent methylation of the phenanthrenediol, which is not isolated, to produce **9,10-DIMETHOXYPHENANTHRENE**. This product is subsequently subjected to Friedel-Crafts acylation to give **3,6-DIACETYL-9,10-DIMETHOXYPHENANTHRENE**. The bis(triisopropylsilyl) enol ether of this diketone undergoes a Diels-Alder reaction with benzoquinone to give a **[7]HELICENEBISQUINONE**, namely 6,13-bis(triisopropylsiloxy)-9,10-dimethoxy[7]-helicenebisquinone, a member of a class of compounds of especial interest as helical ligands, nonlinear optical materials, probes of remote chirality, and so forth.

Not unlike similar, successful organizations, Organic Syntheses continues to evolve. The original Organic Syntheses web site (www.orgsyn.org) has been in existence for nearly two years and a second, Wiley-sponsored site has been created. The membership of both the Board of Editors and Board of Directors undergoes change frequently. One component of this enterprise, however, that has not changed is the all-encompassing role played by Professor Jeremiah Freeman, Secretary to both Boards. Jerry is certainly the single indispensable element of this volume, as he has been for so many preceding volumes. His guidance, encouragement, gentle prodding, and friendship have been invaluable during my tenure as an editor. I will very much miss our collaboration. Regretfully, his retirement with the completion of Volume 81 will deprive subsequent editors the opportunity to discover his many qualities more thoroughly.

I also wish to extend my appreciation for my fellow editors and their students and/or coworkers whose names are cited below for their time-consuming and painstaking efforts in checking the procedures contained herein, ensuring that they are particularly "robust."

Finally, a special thanks is due to the heads of the Chemical Synthesis Department at Hoffmann-La Roche, namely Drs. David L. Coffen and Christopher M. Exon, whose constant support of–indeed, patient indulgence in–my activities as an editor has been essential during these years.

STEVEN WOLFF

Nutley, New Jersey

RICHARD T. ARNOLD
June 18, 1913–January 16, 2001

Richard Thomas Arnold was born on June 18, 1913 in Indianapolis, Indiana. His parents, Mr. and Mrs. Robert H. Arnold, were naturalized American citizens of English origin. He was educated at the Technical High School of Indianapolis through the first year of high school, and then at Collinsville High School in Collinsville, Illinois. He entered Southern Illinois Teachers College (now Southern Illinois University) in Carbondale (1930–34), from which he received a Bachelor of Education degree, and to which he would return in 1970 (see below). He carried out his graduate work at the University of Illinois in Urbana (1934–37), where he received an M.S. degree in 1935 and a Ph.D. in 1937 under the supervision of the late Professor Reynold C. ("Bob") Fuson. Fuson was Editor of Vol. 18 of Organic Syntheses and, as a result, subsequently became a member of the Advisory Board. Besides Arnold, Fuson also served as Ph.D. adviser to six other members of the Advisory Board who also became members (and, in many cases, officers) of the Board of Directors: Norman Rabjohn (Coll. Vol. 4; Treasurer), Blaine

C. McKusick (Vol. 43; President), William E. Parham (Vol. 44; Treasurer, who became a faculty colleague of Arnold at the University of Minnesota), William D. Emmons (Vol. 47; Vice President), Herbert O House (Vol. 52) and Jeremiah P. Freeman (Coll. Vols. 7, 8, 9; Secretary).

Arnold (who was known as "Dick" among his colleagues) joined the faculty of the University of Minnesota as an Instructor of Chemistry in the fall of 1937 at an academic year salary of $2200, following the resignation of Alberto F. Thompson, Jr. Dick married his gracious wife, Doris ("Do") in 1939, and their son, Robert, was born in August 1940, followed by their daughter, Mary Lyn (Wonderlic). Dick was promoted to Assistant Professor in 1940, at which time he had 13 publications, one based on his Ph.D. work and 12 based on his work at Minnesota, largely in the field of physical organic chemistry. In 1940 his first three Ph.D. students graduated: Harold E. Zaugg, J. C. McCool, and Fred J. Bordwell. His reputation grew, and in the words of Lee Irvin Smith on April 15, 1941: "Among chemists he is generally esteemed as the one man who can rank with Paul Bartlett in the field of physico-organic research" (Bartlett in 1932–34 had preceded Arnold at Minnesota). Upon recommendation of three well-known American chemists, none of whom were attached to his alma mater, the University of Illinois, or to the University of Minnesota, Northwestern University made him a handsome offer of $900 more than he was getting at Minnesota. Thus, he became an early "retention case," but with the help of his fondness for Minnesota, and some salary augmentation, he was persuaded to stay.

On August 14, 1942 Arnold and his family moved to 1573 Fulham St., St. Paul, near the University Grove faculty-staff housing area, which remained their home throughout the remainder of his career at the University of Minnesota. During World War II, besides carrying out his normal academic duties, Arnold collaborated for 51% of his time with his senior colleague (and the postdoctoral adviser of this author), Professor Walter M. Lauer, in directing an antimalarial synthesis research project assigned by the Committee on Medical Research of the Office of Scientific Research and Development (O.S.R.D.). He was promoted to Associate Professor in 1944, at which time he had 30 publications. In 1946, as outside offers became more prevalent, he was promoted to Full Professor.

In 1948–49 Arnold became a Guggenheim Memorial Fellow and spent six months at the ETH in Zurich, Switzerland and six months at the Radiation Laboratory in Berkeley, California. In 1949 he was awarded the American Chemical Society (A.C.S.) Award in Pure Chemistry for his detailed explanations of chemical reactions, following in the footsteps of his senior colleague, Professor C. Frederick Koelsch (in 1934, when it was known as the Langmuir Award) and Paul D. Bartlett of Harvard (in 1938).

During the academic year 1950–51 and the fall of 1952, Arnold served as Assistant to the Dean of the Institute of Technology, two-thirds time, while Dean Athelstan F. Spilhaus was engaged in special work on a government research project. After part of a year back full time as a Professor, Arnold took a leave of absence for the academic year 1952–53 to serve as the first Science Attaché to the U.S. High Commissioner in the newly established U.S. Embassy in Bonn, Germany (This vacancy allowed this author to become a member of the faculty at Minnesota). Upon his return to Minnesota in the fall of 1953, Arnold became Head of the Chemistry Department. Less than two years later, after 18 years on the faculty at Minnesota, Arnold resigned, effective June 15, 1955, to become Director of the Basic Research Program in Physical Sciences of the Alfred P. Sloan, Jr., Foundation in New York City. This agency awards the coveted Alfred P. Sloan Fellowships to promising young scientists in the physical sciences.

In 1960, Arnold joined Mead Johnson and Co. in Evansville, Indiana, as Director of Research. Subsequently, he became Vice-President of Research and Development, then President of the Mead Johnson Research Center, and finally Vice-President and Chairman of the Scientific Advisory Board. In 1970, he returned to his undergraduate alma mater, Southern Illinois University, as Professor and Chairman of the Department of Chemistry and Biochemistry. He left the chairmanship in 1975, but remained as Professor until his retirement in 1982. The following year he served as a Visiting Professor at Northwestern University in Evanston. He continued to live in the Evanston area until his death at age 87 on January 16, 2001 in Evanston Hospital, following an accident at the Mather Garden Retirement Home the week before.

Besides the A.C.S. Award in Pure Chemistry received in 1949, Arnold received Honorary D.Sc. Degrees from Northwestern University (1979) and Southern Illinois University. He had over 100 scientific publications. He was noted for his detailed explanations and keen insight into the mechanisms of organic chemical reactions, including his early espousal of "quasi six-membered ring transition states." While at the University of Minnesota, he served as a chemical consultant to five major companies at various times (with the beginning year shown in parentheses): Barrett Division of the Allied Chemical and Dye Corp. (1943), E. I. du Pont de Nemours and Co. (1950), General Mills Research Laboratory (1950–54), Parke, Davis and Co. (1953), and the Procter and Gamble Co. (1955). He was active on departmental and university committees, and as an officer in the Minnesota Section of the A.C.S. and later on National Committees of the Society.

Dick Arnold served an eight-year term on the Board of Editors of Organic Syntheses, and edited Vol. 32, published in 1952. Then, as is customary, he became a member of the Advisory Board and, later, was elected a member of the Board of Directors and served as its Treasurer from 1957–67 and President from 1968 until his retirement from the Board in 1980. He last attended an Organic Syntheses Dinner in Chicago on August 20, 1995, at the age of 82. Dick Arnold will be remembered as a talented researcher, a skillful and enthusiastic teacher, an effective administrator, and an extremely personable individual whom you couldn't help but like. He was preceded in death by his wife, Do, and survived by his son, Robert, and his daughter, Mary Lyn Wonderlic, four grandchildren, and five great-grandchildren.

References

(1) Personnel file in the Chemistry Department at the University of Minnesota.

(2) "Richard T. Arnold," obituary in the Chicago Tribune, Chicago Sports Final, N Edition Jan. 19, 2001, p. 9, ISSN 10856706.

(3) Cal Y. Meyers, "Richard T. Arnold," Chemistry Department, Southern Illinois University, Carbondale, IL, Jan. 24, 2001 (Sent to the A.C.S.)

WAYLAND E. NOLAND

Department of Chemistry
University of Minnesota
Minneapolis, MN 55455

November 17, 2002

WILLIAM D. EMMONS
November 18, 1924–December 8, 2001

William D. Emmons was a native of Minnesota and received his B.S. degree from the University of Minnesota. He served in World War 2 as a meteorologist in the Pacific Theater. After the war he attended graduate school at the University of Illinois, where he received a Ph.D. with R. C. Fuson in 1951.

Facing recall into the service in the Korean War, he opted to join the Redstone Arsenal Research Division of the Rohm and Haas Co., a laboratory being run for the Army Ordnance Corps in the area of solid propellant research. This laboratory grew to be one of the leaders in that area, largely due to Bill's energy and drive. During this period Bill pioneered work in the applications of peroxytrifluoroacetic acid as a powerful oxidant as well as nitration under alkaline conditions. He discovered the long-sought cyclic isomer of nitrones, the oxaziridines, and in a short period of time he explored the remarkable chemistry of these compounds. His review in 1964 consisted of twenty-two pages and 18 references. An updated review in 1985 consisted of sixty-seven pages and 211 references, showing the explosion of interest in these compounds.

In 1957 Rohm and Haas called him to head up a laboratory in the main research center of the company. During this period he introduced with Bill Wadsworth the use of phosphonate esters in Wittig olefination reactions. Bill Emmons always wanted to leave behind a chemical legacy, but he would have found it ironic that his most frequently cited contribution, the phosphonate chemistry, was the one he felt was the least imaginative of his works.

Bill rose through the management levels of Rohm and Haas to the point where his research area, polymers, resins and monomers, was the greatest contributor to the corporate profits. At the same time he always thought of himself primarily as a scientist and, even up to his retirement in 1989, he prided himself on his personal contributions to the research program. In 1986 he received the first Roy W. Tess Award for Coatings Science sponsored by the Division of Polymeric Materials of the American Chemical Society.

His scientific leadership qualities were recognized by the 1993 Earle B. Barnes Award for Leadership in Chemical Research sponsored by the American Chemical Society. The following excerpt from an award nomination by his late colleague, Donald L. Glusker, sums up the man I knew:

"Emmons set a clear example for scientific integrity, both on the facts of experimentation but also in dealing with regulatory issues of toxicity, waste disposal, etc., and he did not tolerate sloppiness or short cuts in this area. His interactions with the young scientists continue, even though he now has almost 300 people in his directorate. He participates intensively in program reviews and continues to put forth effective ideas which turn into major opportunities. He is a master at finding the right spot for wizened old bench chemists to revive their enthusiasm and their flagging careers. Without losing sight of the realities of industrial research in terms of profits, costs, competition, etc., he continues to get most enjoyment from the science of what is being accomplished, and sets a role model for the staff of someone who loves and lives for chemistry."

Bill became an editor of Organic Syntheses in 1961 and served as Editor of Volume 47. He was elected to the Board of Directors in 1971. He served for many years on the Finance Committee of the Board; he retired from the Board in 1995.

Bill had another great passion in life besides chemistry: ballet. He and his wife Margaret commuted regularly from Philadelphia to New York during the ballet season to see performances by the American Ballet Theater and The New York City Ballet.

Another of his great enthusiasms was good food. When I became Secretary of Organic Syntheses, Bill charged me with the task of providing "memorable" meals for our semi-annual dinners. I am not sure that I ever actually met his standard but he was always encouraging me to keep trying.

Last but not least was his passion for his family. He was not a person demonstrative of his feelings. At first acquaintance he might appear brusque and abrupt. But inside a "crusty" exterior was a person of deep and sincere compassion and sensitivity. His wife of many years, Margaret (affectionately known as Quie) and his children, Billy, Jean and Nancy, were the principal beneficiaries of his care and concern in his very private personal life.

I was fortunate and privileged to have been associated early in my career with such a mentor. From my first days as a graduate student in the Fuson group at Illinois in 1950 and then again as a member of Bill's synthesis group at Redstone from 1953 to 1957, I was introduced to the world of chemical research under the aegis of one of the most imaginative chemists of his generation. His standards, his drive, and particularly his ambition, that never came at the expense of others, remain an inspiration for me to this day.

JEREMIAH P. FREEMAN

November 17, 2002

CONTENTS

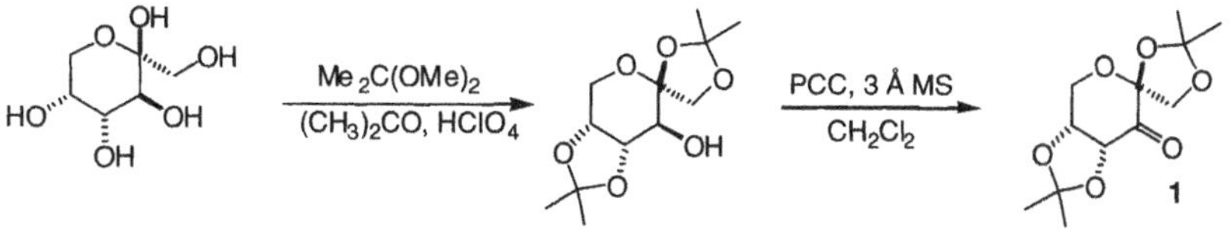

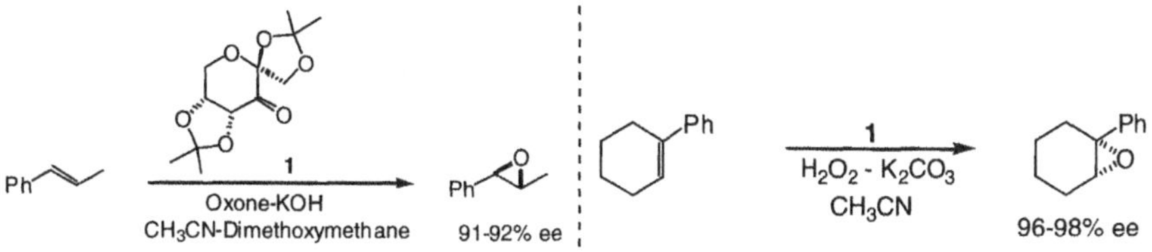

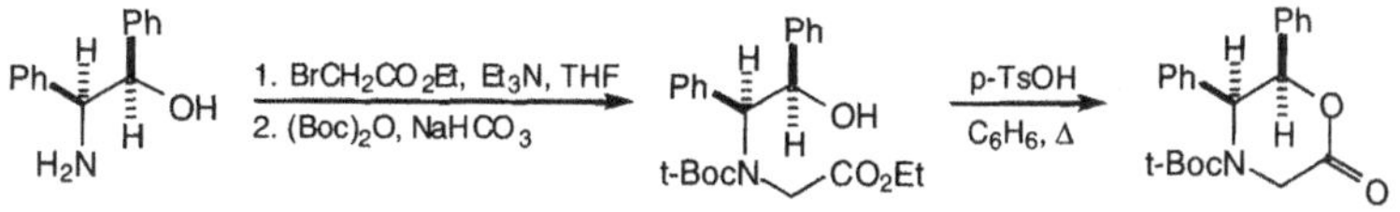

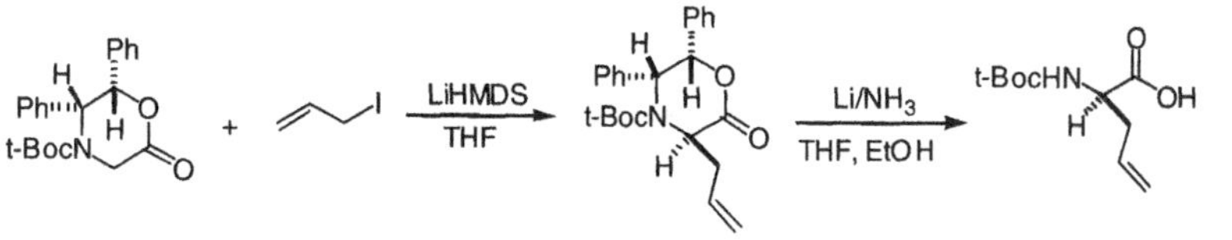

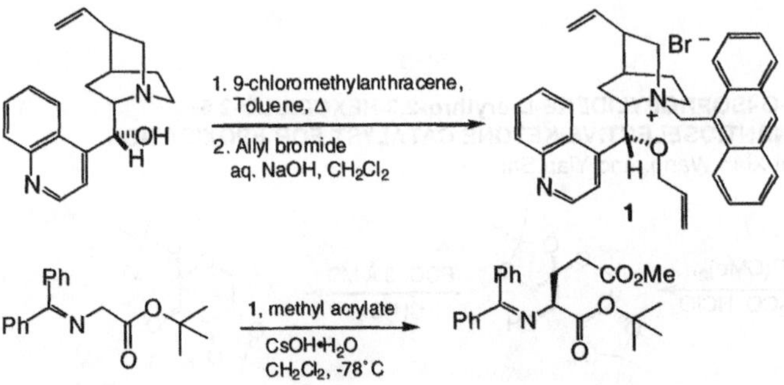
1. 9-chloromethylanthracene, Toluene, Δ
2. Allyl bromide aq. NaOH, CH_2Cl_2
Br⁻
1
1, methyl acrylate
$CsOH{\cdot}H_2O$
CH_2Cl_2, -78°C
Ph
CO_2Me

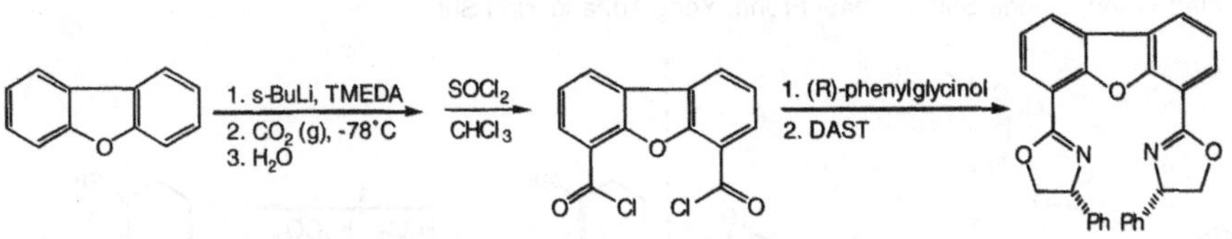
1. s-BuLi, TMEDA
2. CO_2 (g), -78°C
3. H_2O
$SOCl_2$
$CHCl_3$
1. (R)-phenylglycinol
2. DAST

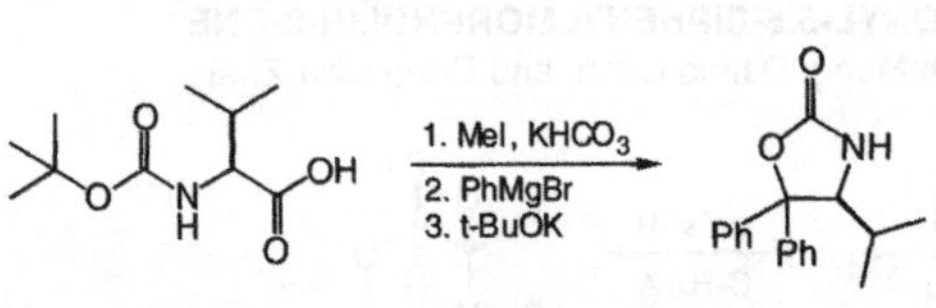
1. MeI, $KHCO_3$
2. PhMgBr
3. t-BuOK

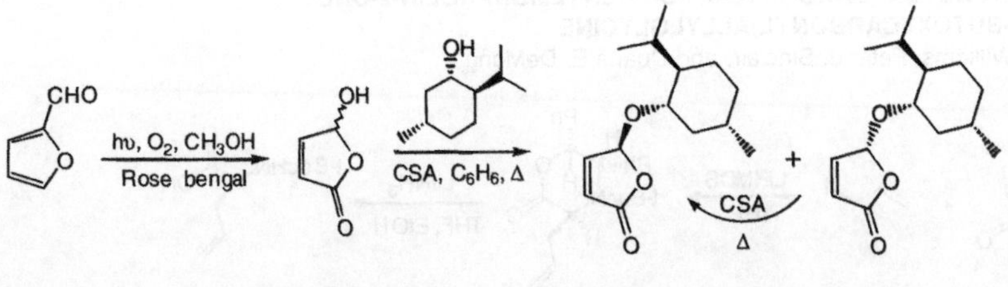
hν, O_2, CH_3OH
Rose bengal
CSA, C_6H_6, Δ
CSA
Δ

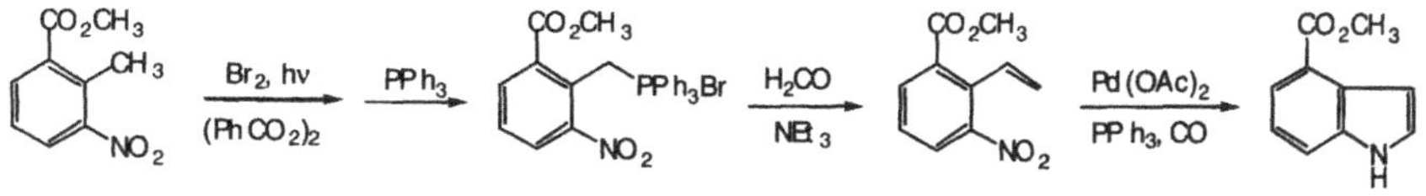

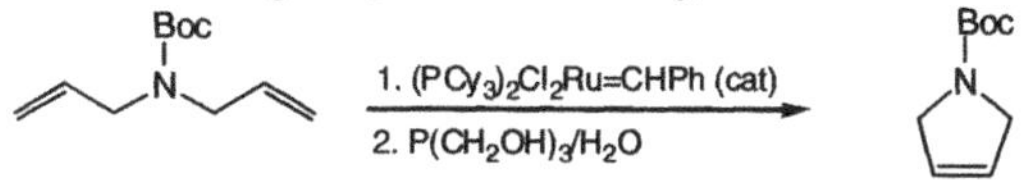

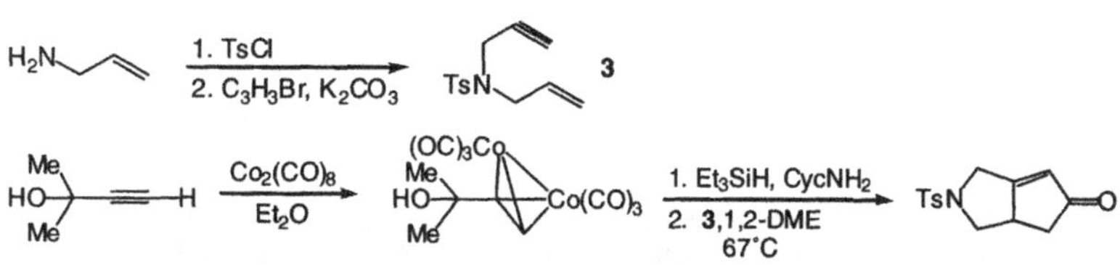

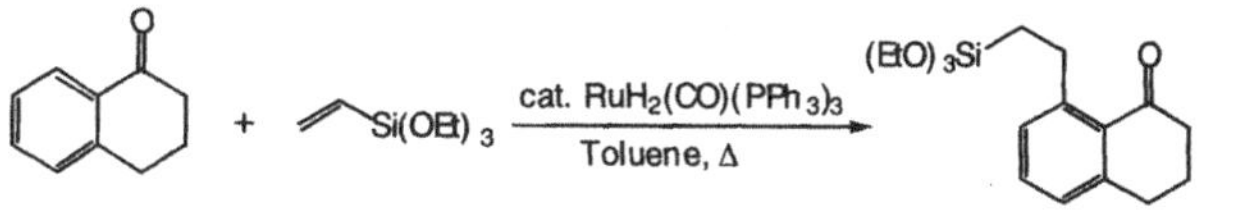

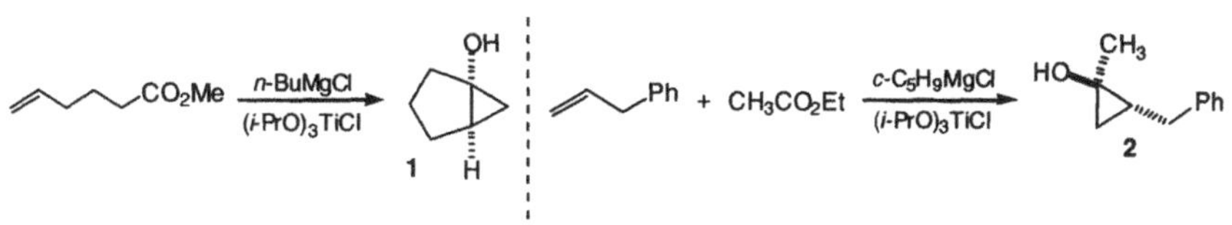

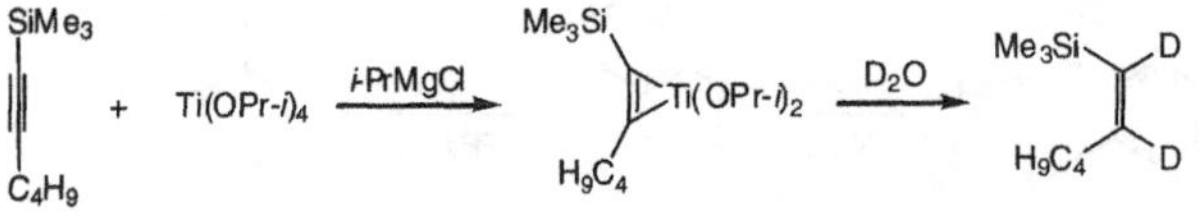

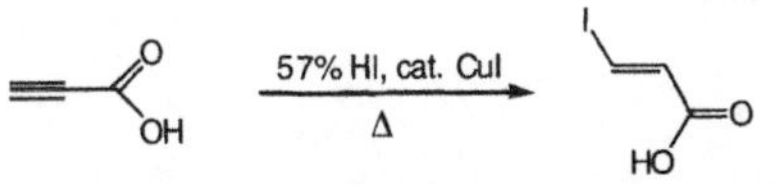

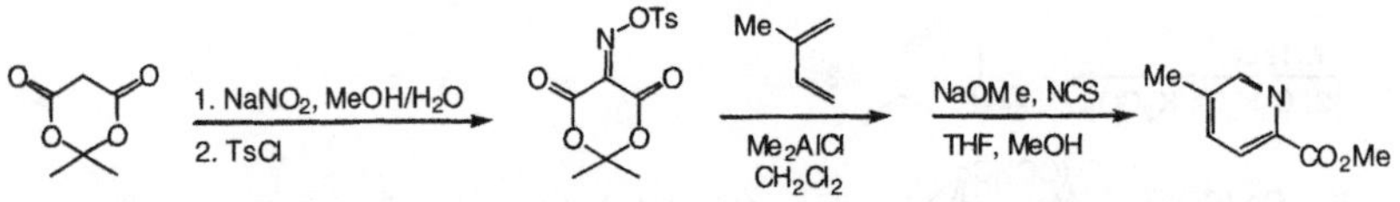

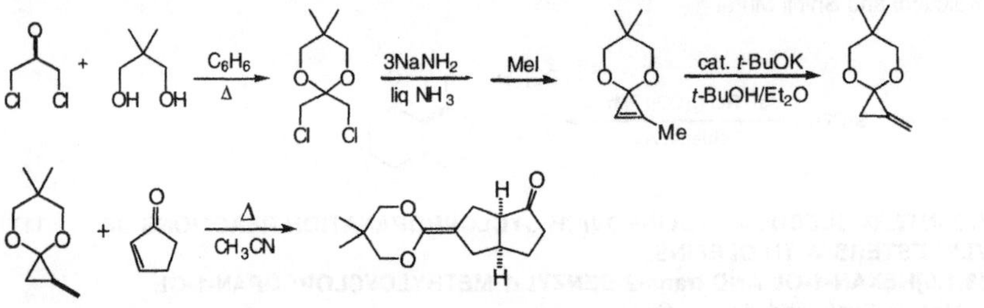

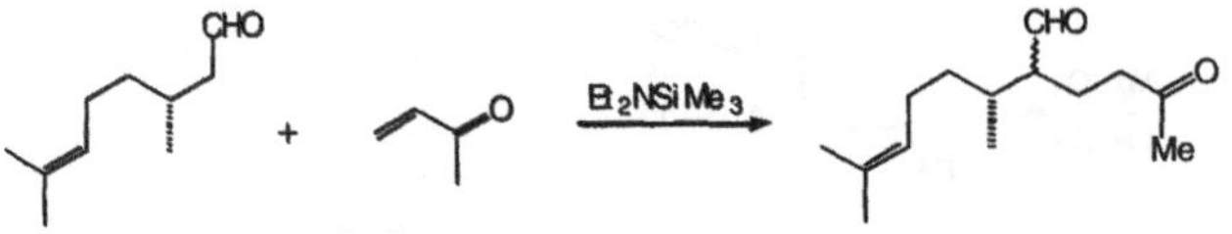

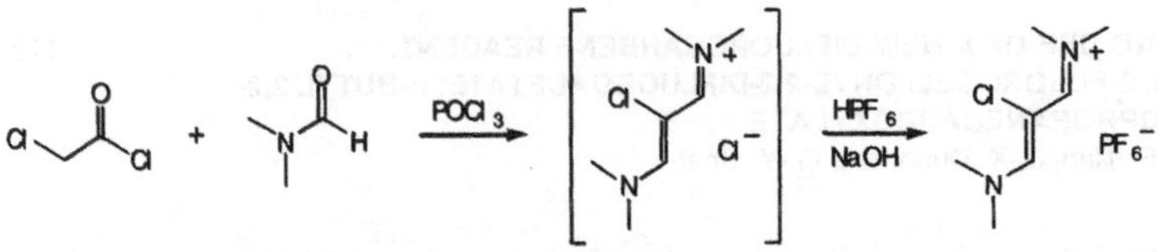

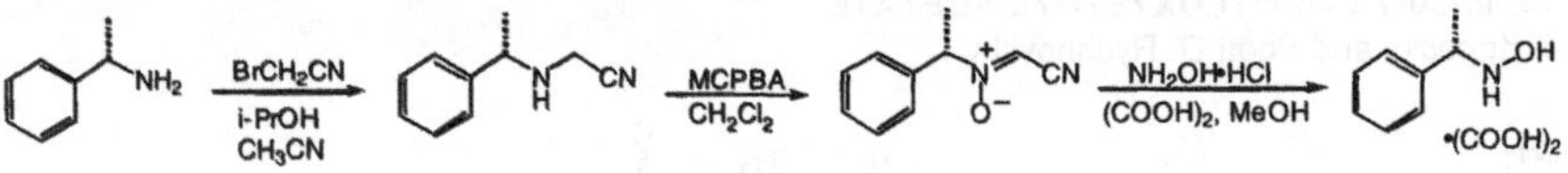

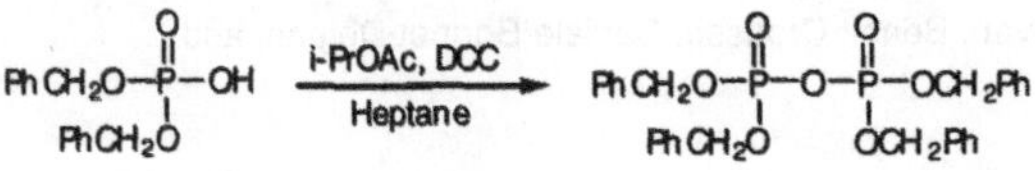

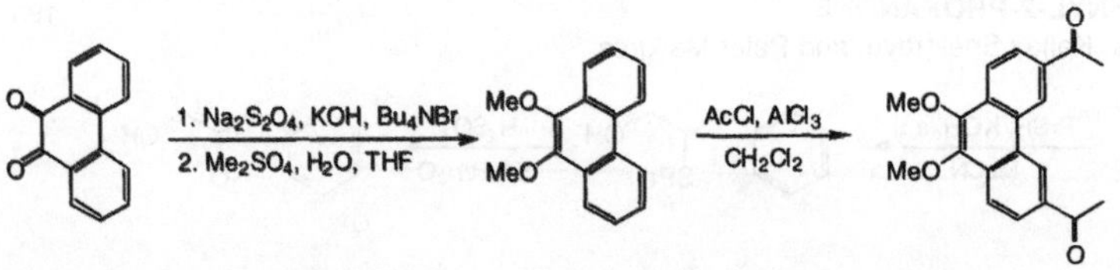

ORGANIC SYNTHESES

SYNTHESIS OF 1,2:4,5-DI-O-ISOPROPYLIDENE-D-erythro-2,3-HEXODIULO-2,6-PYRANOSE. A HIGHLY ENANTIOSELECTIVE KETONE CATALYST FOR EPOXIDATION

[β-D-erythro-2,3-Hexodiulo-2,6-pyranose, 1,2:4,5-bis-O-(1-methylethylidene)-]

A. MeO OMe, $(CH_3)_2CO$, $HClO_4$ → **1**

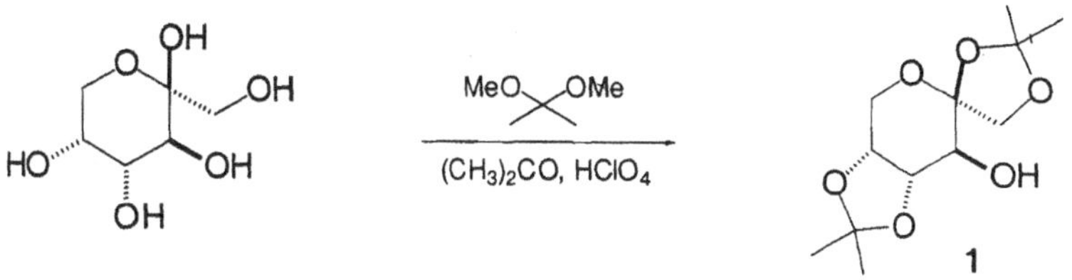

B. **1** → PCC, 3 Å MS, CH_2Cl_2 → **2**

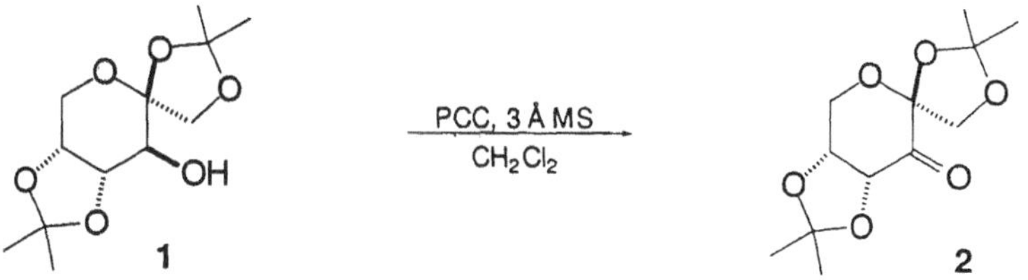

Submitted by Yong Tu, Michael Frohn, Zhi-Xian Wang, and Yian Shi.[1]

Checked by Jason M. Diffendal and Rick L. Danheiser.

1. Procedure

A. *1,2:4,5-Di-O-isopropylidene-β-D-fructopyranose* (**1**).[2] D-Fructose (18.0 g, 100 mmol) (Note 1) and 2,2-dimethoxypropane (7.4 mL, 60 mmol) (Note 1) are added to 350 mL of acetone (Note 1) in a 1-L, round-bottomed flask equipped with a Teflon-coated magnetic stir bar. The flask is cooled in an ice bath for 15 min, then 4.3 mL of 70%

perchloric acid (Note 2) is added in one portion. The resulting suspension is stirred for 6 hr at 0°C (Note 3). Concentrated ammonium hydroxide (4.8 mL) is then added to neutralize the acid and the solvent is removed by rotary evaporation at 25°C to give a white solid. This solid is dissolved in 200 mL of dichloromethane (CH_2Cl_2) and washed with two 50-mL portions of saturated sodium chloride solution, dried over sodium sulfate (Na_2SO_4), filtered, and concentrated by rotary evaporation (25°C) until the total volume is about 40 mL (Note 4). Boiling hexane (100 mL) is then added (Note 5) and the flask is allowed to cool to room temperature, during which time the bulk of the product crystallizes out of solution. Additional product crystallizes upon cooling to –25°C for 4 hr. Isolation of the solid by vacuum filtration and careful washing with three 25-mL portions of cold (–25°C) hexane gives 13.4-13.6 g (51-52%) of the title alcohol as fine white needles (Note 6).

B. *1,2:4,5-Di-O-isopropylidene-D-erythro-2,3-hexodiulo-2,6-pyranose* (**2**).[2] A 500-mL, round-bottomed flask equipped with a 4.5-cm, egg-shaped Teflon-coated magnetic stir bar is charged with 130 mL of CH_2Cl_2 (Note 1), the alcohol prepared in Step A (10.4 g, 40.0 mmol), and 15 g of freshly powdered 3 Å molecular sieves (Note 7). Pyridinium chlorochromate (21.5 g, 100 mmol) (Note 1) is added portionwise over 10 min and the resulting mixture is stirred at room temperature for 15 hr (Note 8). Ether (200 mL) is added slowly with vigorous stirring and the solution is filtered under vacuum through a pad of 35 g of Celite (Note 9). The solids remaining in the reaction flask are transferred to the Celite pad by scraping with a spatula and washing with three 50-mL portions of ether. The resulting cloudy brown filtrate is concentrated by rotary evaporation at room temperature to give a brown solid. To this solid is added 25 mL of 1:1 ether:hexane and the solids are scraped with a spatula. The mixture is then poured onto 60 g of Whatman 60 Å (230-400 mesh) silica gel packed in a 4-cm diameter chromatography column and the liquid is adsorbed onto the silica gel by gravity (Note 10). The material remaining in the flask is further washed with 1:1 ether:hexane and transferred onto the silica gel; this process is repeated until all the material has been loaded onto the silica gel. The ketone is eluted using

500 mL of 1:1 ether:hexane and the eluent is concentrated by rotary evaporation to afford the crude ketone as a white solid. This material is dissolved in 40-45 mL of boiling hexane. Upon cooling the solution to room temperature, the ketone begins to crystallize. The flask is then cooled to –25°C for 2 hr. The resulting solids are collected by filtration, washed with three 25-mL portions of cold (–25°C) hexane, and dried to afford 8.84-9.08 g, (86-88%) of the ketone as a white solid (Note 11).

2. Notes

1. D-Fructose (98%), 2,2-dimethoxypropane (98%), and pyridinium chlorochromate (98%) were obtained from Aldrich Chemical Company, Inc. and used as received. ACS grade acetone and dichloromethane were purchased from Fisher Scientific and used as received.

2. Perchloric acid (70%) was obtained from J. T. Baker Company. Reaction of 70% perchloric acid with organic materials can lead to fires and explosions, and anhydrous $HClO_4$ is potentially explosive. Although no incidents occurred in the experience of the submitters, care should be taken in handling this compound.

3. The suspension turns into a clear, colorless solution after 1-2 hr. The title compound is the kinetic product of the reaction, and can readily isomerize to 2,3:4,5-di-O-isopropylidene-β-D-fructopyranose (the thermodynamic product; see ref. 2c). Control of the reaction time is important to minimize the formation of the thermodynamic product.

4. The solvent volume can vary slightly without much effect on the yield of the recrystallization step.

5. The white, crystalline product begins to precipitate in the first 5 min after the addition of hexane.

6. The product exhibits the following physical and spectral properties: mp 118.5-119.5°C; IR (KBr) cm^{-1}: 3547; ^{1}H NMR (500 MHz, $CDCl_3$) δ: 1.37 (s, 3 H), 1.44 (s, 3 H),

1.52 (s, 3 H), 1.54 (s, 3 H), 1.99 (d, 1 H, J = 8.1), 3.67 (dd, 1 H, J = 8.1, 6.8), 3.98 (d, 1 H, J = 9.0), 4.01 (dd, 1 H, J = 13.2, 0.9), 4.12 (dd, 1 H, J = 13.2, 2.7), 4.13 (dd, 1 H, J = 6.8, 5.7), 4.19 (d, 1 H, J = 9.0), 4.22 (ddd, 1 H, J = 5.7, 2.7, 0.9); ^{13}C NMR (125 MHz, $CDCl_3$) δ: 26.1, 26.5, 26.6, 28.1, 61.0, 70.6, 73.52, 73.53, 77.5, 104.7, 109.6, 112.0; HRMS (ESI) Calcd for $C_{12}H_{20}O_6$ $[M+Na]^+$: 283.1152; Found: 283.1149; $[\alpha]_D^{25}$-144.2 ($CHCl_3$, c 1.0).

7. Mallinckrodt Grade 564 CCGT3A molecular sieves are used without further activation.

8. The mixture turns from orange-brown to a dark brown color during the course of the reaction, indicating the reduction of Cr(VI) to Cr(III).

9. Slow addition of ether is necessary for a high yield. The addition of ether leads to the precipitation of only a small amount of the reduced chromium. This filtration mainly removes the molecular sieves and chromium species adsorbed during stirring.

10. The yield of the ketone will be reduced if all the brown solids are not loaded onto the column (these solids contain some of the ketone).

11. The product exhibits the following physical and spectral properties: mp 101-103°C; IR (KBr) cm^{-1}: 1749; ^{1}H NMR (500 MHz, $CDCl_3$) δ: 1.40 (s, 6 H), 1.46 (s, 3 H), 1.55 (s, 3 H), 4.00 (d, 1 H, J = 9.5), 4.12 (d, 1 H, J = 13.4), 4.39 (dd, 1 H, J = 13.4, 2.2), 4.55 (ddd, 1 H, J = 5.7, 2.2, 1.0), 4.61 (d, 1 H, J = 9.5), 4.73 (d, 1 H, J = 5.7); ^{13}C NMR (125 MHz, $CDCl_3$) δ: 26.20, 26.24, 26.7, 27.3, 60.3, 70.2, 76.1, 78.1, 104.3, 110.8, 114.0, 197.1; HRMS (ESI) Calcd for $C_{12}H_{18}O_6$ $[M+Na]^+$: 281.0996; Found: 281.0985; $[\alpha]_D^{25}$−125.4 ($CHCl_3$, c 1.0).

Waste Disposal Information

All toxic materials were disposed of in accordance with "Prudent Practices in the Laboratory"; National Academy Press; Washington, DC, 1995.

3. Discussion

Dioxiranes are remarkably versatile oxidizing agents which show encouraging potential for asymmetric synthesis, particularly asymmetric epoxidation. Dioxiranes can be generated in situ from Oxone ($KHSO_5$) and ketones (Scheme 1).[3] In principle, only a catalytic amount of ketone is required, so with a chiral ketone there exists the opportunity for catalytic asymmetric epoxidation.[4,5] Since the first asymmetric epoxidation of olefins with a chiral dioxirane reported by Curci in 1984,[4a] this area has received intensive interest and significant progress has been made.[4,5]

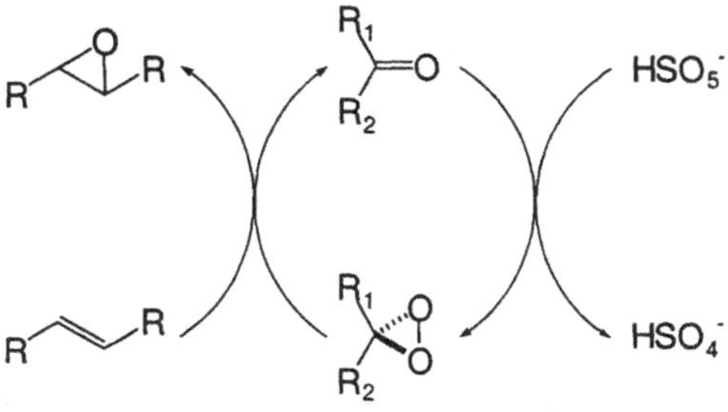

Scheme 1

The fructose-derived ketone (**2**) described herein is a member of a class of ketones designed to contain the following general features: (1) The stereogenic centers are close to the reacting center, resulting in efficient transfer of stereochemistry between substrate and catalyst. (2) The presence of fused ring(s) or a quaternary center α to the carbonyl group minimizes epimerization of the stereogenic centers. (3) Approach of an olefin to the reacting dioxirane can be controlled by sterically blocking one face or using a C_2- or pseudo-C_2-symmetric element. (4) Electron-withdrawing (by induction) substitutents are introduced to activate the carbonyl.

Ketone **2** gives very high enantioselectivities for a variety of trans-disubstituted and trisubstituted olefins.[5] The ketone catalyst can be readily synthesized from very inexpensive D-fructose by ketalization with acetone[2e,h] and subsequent oxidation of the remaining alcohol to the ketone. Other acids such as H_2SO_4 can also be used for ketalization.[2c,f,i] Although the present procedure uses PCC for the oxidation, many other oxidants such as PDC-Ac_2O,[2f] DMSO-Ac_2O,[2a,b,d] DMSO-DCC,[2e] DMSO-$(COCl_2)$,[2g] $RuCl_3$-$NaIO_4$,[2h] Ru-TBHP,[2j] etc. are also effective. The enantiomer of catalyst **2** **(ent-2)** can be prepared in the same fashion from L-fructose, which in turn can be prepared from readily available L-sorbose.[6,5c] As expected, the enantiomeric catalyst shows the same enantioselectivity in epoxidation reactions.

1. Department of Chemistry, Colorado State University, Fort Collins, CO 80523.

2. For leading references see: (a) McDonald, E. J. *Carbohydr. Res.* **1967**, *5*, 106; (b) James, K.; Tatchell, A. R.; Ray, P. K. *J. Chem. Soc. C* **1967**, 2681; (c) Brady, R. F., Jr. *Carbohydr. Res.* **1970**, *15*, 35; (d) Tipson, R. S.; Brady, R. F. Jr.; West, B. F. *Carbohydr. Res.* **1971**, *16*, 383; (e) Prisbe, E. J.; Smejkal, J.; Verheyden, J. P. H.; Moffatt, J. G. *J. Org. Chem.*, **1976**, *41*, 1836; (f) Fayet, C.; Gelas, J. *Carbohydr. Res.* **1986**, *155*, 99; (g) Maryanoff, B. E.; Reitz, A. B.; Nortey, S. O. *Tetrahedron* **1988**, *44*, 3093; (h) Mio, S.; Kumagawa, Y.; Sugai, S. *Tetrahedron* **1991**, *47*, 2133; (i) Kang, J.; Lim, G. J.; Yoon, S. K.; Kim, M. Y. *J. Org. Chem.* **1995**, *60*, 564; (j) Fung, W.-H.; Yu, W.-Y.; Che, C.-M. *J. Org. Chem.* **1998**, *63*, 2873.

3. For general leading references on dioxiranes see: (a) Adam, W.; Curci, R.; Edwards, J. O. *Acc. Chem. Res.* **1989**, *22*, 205; (b) Murray, R. W. *Chem. Rev.* **1989**, *89*, 1187; (c) Curci, R.; Dinoi, A.; Rubino, M. F. *Pure & Appl. Chem.* **1995**, *67*, 811; (d) Clennan, E. L. *Trends Org. Chem.* **1995**, *5*, 231; (e) Adam, W.; Smerz, A. K. *Bull. Soc. Chim. Belg.* **1996**, *105*, 581; (f) Denmark, S. E.; Wu, Z. *Synlett* **1999**, 847.
4. For leading references on asymmetric epoxidation mediated by chiral ketones see: (a) Curci, R.; Fiorentino, M.; Serio, M. R. *J. Chem. Soc., Chem. Commun.* **1984**, 155; (b) Curci, R.; D'Accolti, L.; Fiorentino, M.; Rosa, A. *Tetrahedron Lett.* **1995**, *36*, 5831; (c) Denmark, S. E.; Forbes, D. C.; Hays, D. S.; DePue, J. S.; Wilde, R. G. *J. Org. Chem.* **1995**, *60*, 1391; (d) Brown, D. S.; Marples, B. A.; Smith, P.; Walton, L. *Tetrahedron* **1995**, *51*, 3587; (e) Yang, D.; Yip, Y.-C.; Tang, M.-W.; Wong, M.-K.; Zheng, J.-H.; Cheung, K.-K. *J. Am. Chem. Soc.* **1996**, *118*, 491; (f) Yang, D.; Wang, X.-C.; Wong, M.-K.; Yip, Y.-C.; Tang, M.-W. *J. Am. Chem. Soc.* **1996**, *118*, 11311; (g) Song, C. E.; Kim, Y. H.; Lee, K. C.; Lee, S.; Jin, B. W. *Tetrahedron: Asymmetry* **1997**, *8*, 2921; (h) Adam, W.; Zhao, C.-G. *Tetrahedron: Asymmetry* **1997**, *8*, 3995; (i) Denmark, S. E.; Wu, Z.; Crudden, C. M.; Matsuhashi, H. *J. Org. Chem.* **1997**, *62*, 8288; (j) Wang, Z.-X.; Shi, Y. *J. Org. Chem.* **1997**, *62*, 8622; (k) Armstrong, A.; Hayter, B. R. *Chem. Commun.* **1998**, 621; (l) Yang, D.; Wong, M.-K.; Yip, Y.-C.; Wang, X.-C; Tang, M.-W.; Zheng, J.-H.; Cheung, K.-K. *J. Am. Chem. Soc.* **1998**, *120*, 5943; (m) Yang, D.; Yip, Y.-C.; Chen, J.; Cheung, K.-K. *J. Am. Chem. Soc.* **1998**, *120*, 7659; (n) Adam, W.; Saha-Möller, C. R.; Zhao, C.-G. *Tetrahedron: Asymmetry* **1999**, *10*, 2749; (o) Wang, Z.-X.; Miller, S. M.; Anderson, O. P.; Shi, Y. *J. Org. Chem.* **1999**, *64*, 6443.
5. For examples of asymmetric epoxidation mediated by fructose-derived ketones see: (a) Tu, Y.; Wang, Z.-X.; Shi, Y. *J. Am. Chem. Soc.* **1996**, *118*, 9806; (b) Wang, Z.-X.; Tu, Y.; Frohn, M.; Shi, Y. *J. Org. Chem.* **1997**, *62*, 2328; (c) Wang,

Z.-X.; Tu, Y.; Frohn, M.; Zhang, J.-R.; Shi, Y. *J. Am. Chem. Soc.* **1997**, *119*, 11224; (d) Frohn, M.; Dalkiewicz, M.; Tu, Y.; Wang, Z.-X.; Shi, Y. *J. Org. Chem.* **1998**, *63*, 2948; (e) Wang, Z.-X.; Shi, Y. *J. Org. Chem.* **1998**, *63*, 3099; (f) Cao, G.-A.; Wang, Z.-X.; Tu, Y.; Shi, Y. *Tetrahedron Lett.* **1998**, *39*, 4425; (g) Zhu, Y.; Tu, Y.; Yu, H.; Shi, Y. *Tetrahedron Lett.* **1998**, *39*, 7819; (h) Tu, Y.; Wang, Z.-X.; Frohn, M.; He, M.; Yu, H.; Tang, Y.; Shi, Y. *J. Org. Chem.* **1998**, *63*, 8475; (i) Wang, Z.-X.; Cao, G.-A.; Shi, Y. *J. Org. Chem.* **1999**, *64*, 7646; (j) Warren, J. D.; Shi, Y. *J. Org. Chem.* **1999**, *64*, 7675; (k) Shu, L.; Shi, Y. *Tetrahedron Lett.* **1999**, *40*, 8721.

6. Chen, C.-C.; Whistler, R. L. *Carbohydr. Res.* **1988**, *175*, 265.

Appendix

Chemical Abstracts Nomenclature (Collective Index Number); (Registry Number)

1,2:4,5-Di-O-isopropylidene-D-erythro-2,3-hexodiulo-2,6-pyranose: β-D-erythro-2,3-Hexodiulo-2,6-pyranose, 1,2:4,5-bis-O-(1-methylethylidene)- (9); (18422-53-2)

D-Fructose (9); (57-48-7)

2,2-Dimethoxypropane: Propane, 2,2-dimethoxy- (9); (77-76-9)

Perchloric acid (8, 9); (7601-90-3)

1,2:4,5-Di-O-isopropylidene-β-D-fructopyranose: β-D-Fructopyranose, 1,2;4,5-bis-O-(1-methylethylidene)- (9); (25018-67-1)

Pyridinium chlorochromate: Chromate(1-), chlorotrioxo-, (T-4)-, hydrogen compd. with pyridine (1:1) (9); (26299-14-9)

ASYMMETRIC EPOXIDATION OF trans-β-METHYLSTYRENE AND 1-PHENYLCYCLOHEXENE USING A D-FRUCTOSE-DERIVED KETONE: (R,R)-trans-β-METHYLSTYRENE OXIDE AND (R,R)-1-PHENYLCYCLOHEXENE OXIDE

(Oxirane, 2-methyl-3-phenyl-, (2R,3R)- and 7-Oxabicyclo[4.1.0]heptane, 1-phenyl-)

A. Ph-CH=CH-CH3 → (1; Oxone-KOH; CH_3CN-Dimethoxymethane) → Ph-epoxide, 91-92% ee

B. 1-phenylcyclohexene → (1; H_2O_2 - K_2CO_3; CH_3CN) → 1-phenylcyclohexene oxide, 96-98% ee

Submitted by Zhi-Xian Wang, Lianhe Shu, Michael Frohn, Yong Tu, and Yian Shi.[1]

Checked by Jason M. Diffendal and Rick L. Danheiser.

1. Procedure

A. *(R,R)-trans-β-Methylstyrene oxide.*[2] A 2-L, three-necked, round-bottomed flask (Note 1) equipped with a 5-cm, egg-shaped, Teflon-coated stir bar and two addition funnels is cooled in an ice-bath. The flask is charged with trans-β-methylstyrene (5.91 g.

50.0 mmol) (Note 2), 500 mL of a 2:1 mixture of dimethoxymethane (Note 2) and acetonitrile (CH_3CN) (Note 2), 300 mL of potassium carbonate-acetic acid buffer solution (Note 3), tetrabutylammonium hydrogen sulfate (0.375 g, 1.1 mmol), and the chiral ketone **1** (4.52 g, 17.5 mmol, 35 mol%) (Note 4). One addition funnel is charged with a solution of Oxone (46.1 g, 75.0 mmol) in 170 mL of aqueous 4×10^{-4}M disodium ethylenediaminetetraacetate (Na_2EDTA) solution (Note 5) and the other addition funnel is charged with 170 mL of 1.47M aqueous potassium hydroxide (KOH) solution. The two solutions in the addition funnels are added dropwise at the same rate over 2.5 hr to the cooled reaction mixture which is stirred vigorously at 0°C (Notes 6, 7). The resulting suspension is stirred at 0°C for an additional hour and then 250 mL of pentane is added. The aqueous phase is separated and extracted with two 250-mL portions of pentane, and the combined organic phases are dried over Na_2SO_4, filtered, and concentrated by rotary evaporation at 0°C (Note 8). The resulting oil is loaded onto 50 g of Whatman 60 Å (230-400 mesh) silica gel (Note 9) packed in a 5-cm diameter column. The silica gel is first washed with 200 mL of hexane to remove trace amounts of unreacted olefin, then the product is eluted with 200 mL of 10:1 hexane:ether to afford 6.02-6.31 g (90-94%) of trans-β-methylstyrene oxide (Notes 10-12).

B. *(R,R)-1-Phenylcyclohexene oxide.*[3] A 250-mL, round-bottomed flask (Note 1) equipped with a 4.5-cm, egg-shaped Teflon-coated magnetic stir bar is charged with 1-phenylcyclohexene (7.91 g, 50.0 mmol) (Note 2) and the chiral ketone **1** (1.29 g, 5.00 mmol, 10 mol%) (Note 4). The flask is cooled in an ice-bath, and 75 mL of CH_3CN and 75 mL of a solution which is 2.0M potassium carbonate and 4×10^{-4}M EDTA are added. The reaction mixture is cooled to 0°C, and 20 mL (200 mmol) of 30% hydrogen peroxide (H_2O_2) is added. The resulting mixture is vigorously stirred at 0°C for 6 hr (Note 13), then diluted with 50 mL of hexane. The aqueous phase is separated and extracted with three 200-mL portions of hexane, and the combined organic phases are washed with two 50-mL portions of 1M aqueous sodium thiosulfate solution and 100 mL of brine, dried over Na_2SO_4, filtered, and concentrated by rotary evaporation at 0°C. The resulting oil is applied to 180 g of Whatman 60 Å (230-400 mesh) silica gel (Note 9) packed in a 5-cm diameter column, then the product is eluted with 600 mL of hexane and finally 1 L of 20:1 hexane:Et_2O to afford 6.88-8.01 g (79-92%) of (R,R)-1-phenylcyclohexene oxide as a colorless oil (Notes 14, 15).

2. Notes

1. All glassware used for the epoxidation reaction is carefully washed to remove trace metals, which may catalyze the decomposition of Oxone and H_2O_2. The checkers used Alconox, followed by water, and then acetone.

2. β-Methylstyrene (99%) and dimethoxymethane (99%) were obtained from Aldrich Chemical Company, Inc. and used as received. ACS grade acetonitrile and 30% H_2O_2 were purchased from Fisher Scientific and used as received. 1-Phenylcyclohexene (99%) was obtained from Alfa Aesar.

3. The buffer solution is prepared by adding 4.5 mL of glacial acetic acid (Fisher Scientific) to 1 L of a 0.1M solution of K_2CO_3 (Fisher Scientific).

4. The chiral ketone **1** was prepared as described in *Org. Synth.* **2003,** *80,* 1.

5. Oxone was obtained from the Aldrich Chemical Company, Inc. The activity of commercial Oxone in oxidation reactions occasionally varies with different batches. Na_2EDTA was purchased from Fisher Scientific.

6. The concentration of Oxone in the reaction mixture and the reaction pH are very important factors in determining the epoxidation efficiency. Both the Oxone and KOH solutions must be added to the reaction mixture in a steady, uniform manner over 2.5 hr.

7. As the reaction progresses, the organic and aqueous phases separate. Salts

precipitate during the first 10-20 min of addition. Vigorous stirring is required to sufficiently mix the two phases; however, excessive splashing of the reaction mixture must be avoided in order to maximize conversion.

8. The epoxide product is volatile. Care should be taken to minimize the loss of the epoxide during concentration.

9. The silica gel is buffered by packing the column with hexane containing 1% triethylamine.

10. The product exhibits the following physical and spectral properties: ^{1}H NMR (500 MHz, $CDCl_3$) δ: 1.45 (d, 3 H, J = 5.1), 3.03 (qd, 1 H, J = 5.1, 2.1), 3.57 (d, 1 H, J = 2.1), 7.20-7.40 (m, 5 H); ^{13}C NMR (125 MHz, $CDCl_3$) δ: 18.1, 59.2, 59.7, 125.7, 128.2, 128.6, 137.9; Anal. Calcd C, 80.56; H, 7.51. Found: C, 80.39; H, 7.37; $[\alpha]_D^{25}$ +45.3 ($CHCl_3$, c 1.84).

11. The submitters obtained the product in 91-92% ee as determined by chiral GC with a Chiraldex G-TA column (25 μm x 30 m) (oven temperature: 80°C; head pressure: 20 psi; retention time: minor isomer at 11.8 min, major isomer at 15.8 min). The checkers obtained epoxide with 89% ee.

12. The submitters report that the ee can be increased to 94% if the epoxidation is carried out at −8 to −10°C.

13. The reaction was monitored by TLC and was complete after 6 hr. The reaction rate is affected by the rate of stirring.

14. The product exhibits the following physical and spectral properties: ^{1}H NMR (500 MHz, $CDCl_3$) δ: 1.25-1.65 (m, 4 H), 1.95-2.03 (m, 2 H), 2.12 (dt, 1 H, J = 15.0, 5.1), 2.29 (ddd, 1 H, J = 15.0, 8.4, 5.1), 3.09 (br s, 1 H), 7.20-7.40 (m, 5 H); ^{13}C NMR (125 MHz, $CDCl_3$) δ: 19.8, 20.1, 24.7, 28.9, 60.2, 61.9, 125.3, 127.1, 128.2, 142.5. Anal. Calcd: C, 82.72; H, 8.10. Found: C, 82.76; H, 8.13; $[\alpha]_D^{25}$ +113.0 (benzene, c 0.56).

15. The submitters obtained the product in 96-98% ee as determined by chiral GC with a Chiraldex G-TA column (25 μm x 30 m) (oven temperature: 80°C; head pressure: 25 psi; retention time: minor isomer at 68.0 min, major isomer at 71.8 min). The checkers obtained the product in 94-95% ee.

Waste Disposal Information

All toxic materials were disposed of in accordance with "Prudent Practices in the Laboratory"; National Academy Press; Washington, DC, 1995.

3. Discussion

The epoxidation procedure described herein utilizes the fructose-derived ketone (**1**) as catalyst and Oxone[4a-j] or H_2O_2[4k,l] as oxidant. The procedure provides a valuable method for the preparation of enantiomerically-enriched epoxides from trans- and trisubstituted olefins. High enantioselectivies have been obtained for a variety of unfunctionalized trans-disubstituted and trisubstituted olefins,[4a-c] vinylsilanes,[4j] hydroxyalkenes,[4e] conjugated dienes,[4d] conjugated enynes,[4f,i] and enol derivatives.[4g] Representative examples are shown in Table I.

Previously, the generation of dioxiranes almost exclusively used potassium peroxomonosulfate ($KHSO_5$) as oxidant.[5,6] Recently, we found that hydrogen peroxide (H_2O_2) could be used as primary oxidant in combination with a nitrile for the epoxidation catalyzed by **1**.[4k,l] In this epoxidation, the peroxyimidic acid generated from the addition of H_2O_2 to CH_3CN is likely to be the active oxidant for the formation of the dioxirane.[7] The Oxone procedure described herein can be applied to a wide variety of olefins without the need for extensive optimization. The more recent H_2O_2 procedure uses much less solvent and salts and is operationally simpler. However, this procedure is somewhat sensitive to the reactivity and solubility of olefins; some optimization (varying catalyst loading, reaction time, reaction temperature, amount of H_2O_2 and solvent) may be required for different substrates.[4l]

Table 1. Asymmetric Epoxidation of Representative Olefins by Ketone **1** (0.3 equiv) with Oxone

Entry	Substrate	Product	Yield (%)	ee (%)	Ref.
1.			85	98	4c
2.	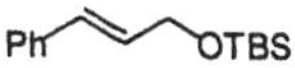		87	94	4c
3.	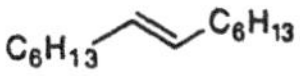		89	95	4c
4.	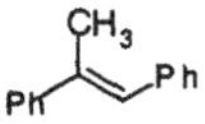	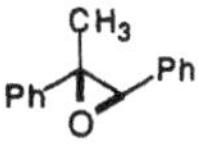	89	95	4c
5.	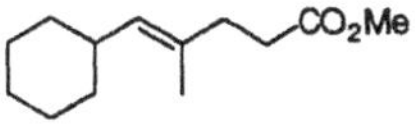	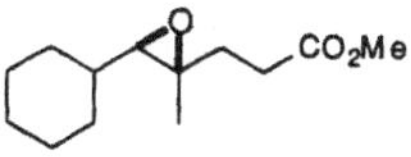	89	94	4c
6.	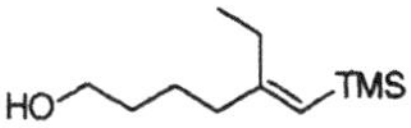	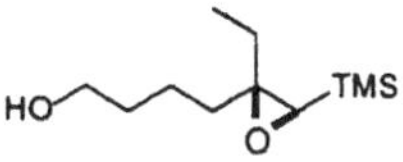	71	93	4j
7.			85	94	4e
8.	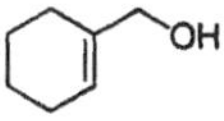	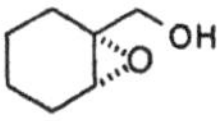	93	94	4e
9.		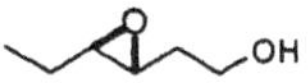	82	90	4e
10.	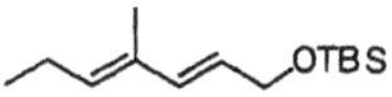	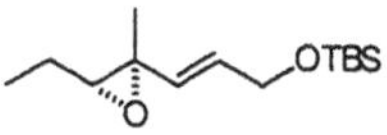	81	96	4d
11.	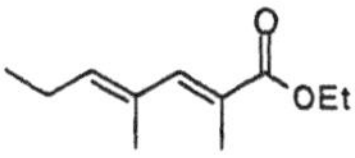	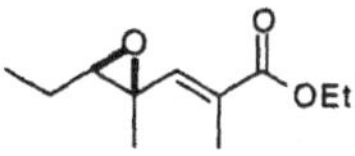	89	94	4d
12.	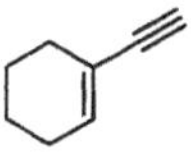	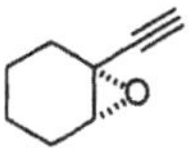	78	93	4i
13.	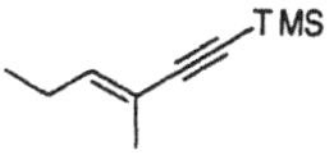		84	95	4i
14.	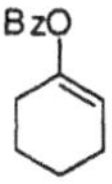	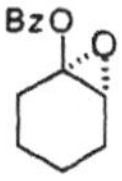	82	93	4g
15.	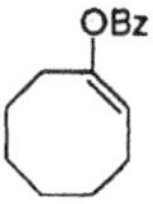	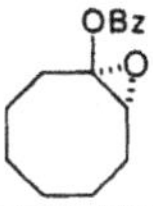	82	95	4g

1. Department of Chemistry, Colorado State University, Fort Collins, CO 80523.
2. Witkop, B.; Foltz, C. M. *J. Am. Chem. Soc.* **1957**, *79*, 197.
3. Berti, G.; Macchia, B.; Macchia, F.; Monti, L. *J. Org. Chem.* **1968**, *33*, 4045.
4. For examples of asymmetric epoxidation mediated by fructose-derived ketones see: (a) Tu, Y.; Wang, Z.-X.; Shi, Y. *J. Am. Chem. Soc.* **1996**, *118*, 9806; (b) Wang, Z.-X.; Tu, Y.; Frohn, M.; Shi, Y. *J. Org. Chem.* **1997**, *62*, 2328; (c) Wang, Z.-X.; Tu, Y.; Frohn, M.; Zhang, J.-R.; Shi, Y. *J. Am. Chem. Soc.* **1997**, *119*, 11224; (d) Frohn, M.; Dalkiewicz, M.; Tu, Y.; Wang, Z.-X.; Shi, Y. *J. Org. Chem.* **1998**, *63*, 2948; (e) Wang, Z.-X.; Shi, Y. *J. Org. Chem.* **1998**, *63*, 3099; (f) Cao, G.-A.; Wang, Z.-X.; Tu, Y.; Shi, Y. *Tetrahedron Lett.* **1998**, *39*, 4425; (g) Zhu, Y.; Tu, Y.; Yu, H.; Shi, Y. *Tetrahedron Lett.* **1998**, *39*, 7819; (h) Tu, Y.; Wang, Z.-X.; Frohn, M.; He, M.; Yu, H.; Tang, Y.; Shi, Y. *J. Org. Chem.* **1998**, *63*, 8475; (i) Wang, Z.-X.; Cao, G.-A.; Shi, Y. *J. Org. Chem.* **1999**, *64*, 7646; (j) Warren, J. D.; Shi, Y. *J. Org. Chem.* **1999**, *64*, 7675; (k) Shu, L.; Shi, Y. *Tetrahedron Lett.* **1999**, *40*, 8721; (l) Shu, L.; Shi, Y. *Tetrahedron* **2001**, 57, 5213.
5. Oxone ($2KHSO_5 \cdot KHSO_4 \cdot K_2SO_4$) is currently the common source of potassium peroxomonosulfate ($KHSO_5$).
6. As close analogues of potassium peroxomonosulfate, arenesulfonic peracids generated from (arenesulfonyl)imidazole/H_2O_2/NaOH have also been shown to produce dioxiranes from acetone and trifluoroacetone as illustrated by ^{18}O-labeling experiments see: Schulz, M.; Liebsch, S.; Kluge, R.; Adam, W. *J. Org. Chem.* **1997**, *62*, 188.
7. For leading references on epoxidation using H_2O_2 and RCN see: (a) Payne, G. B.; Deming, P. H.; Williams, P. H. *J. Org. Chem.* **1961**, *26*, 659; (b) Payne, G. B. *Tetrahedron* **1962**, *18*, 763; (c) McIsaac, J. E. Jr.; Ball, R. E.; Behrman, E. J. *J. Org.*

Chem. **1971**, *36*, 3048; (d) Bach, R. D.; Knight, J. W. *Org. Synth. Coll. Vol. VII* **1990**, 126; (e) Arias, L. A.; Adkins, S.; Nagel, C. J.; Bach, R. D. *J. Org. Chem.* **1983**, *48*, 888.

Appendix
Chemical Abstracts Nomenclature (Collective Index Number); (Registry Number)

(R,R)-trans-β-Methylstyrene oxide: Oxirane, 2-methyl-3-phenyl-, (2R,3R)- (9); (14212-54-5)

trans-β-Methylstyrene: Benzene, (1E)-1-propenyl- (9); (873-66-5)

Dimethoxymethane: Methane, dimethoxy- (8, 9); (109-87-5)

1,2:4,5-Di-O-isopropylidene-D-erythro-2,3-hexodiulo-2,6-pyranose: β-D-erythro-2,3-Hexodiulo-2,6-pyranose, 1,2:4,5-bis-O-(1-methylethylidene)- (9); (18422-53-2)

Tetrabutylammonium hydrogen sulfate: 1-Butanaminium, N,N,N-tributyl-, sulfate (1:1) (9); (32503-27-8)

Disodium ethylenediaminetetraacetate: Glycine, N,N'-1,2-ethanediylbis [N-(carboxymethyl)-, disodium salt (9); (139-33-3)

(R,R)-1-Phenylcyclohexene oxide: 7-Oxabicyclo[4.1.0]heptane, 1-phenyl-, (1R,6R)- (9); (17540-04-4)

Oxone: Peroxymonosulfuric acid, monopotassium salt, mixt. with dipotassium sulfate and potassium hydrogen sulfate (9); (37222-66-5)

ASYMMETRIC SYNTHESIS OF N-tert-BUTOXYCARBONYL α-AMINO ACIDS. SYNTHESIS OF (5S,6R)-4-tert-BUTOXYCARBONYL-5,6-DIPHENYLMORPHOLIN-2-ONE

(4-Morpholinecarboxylic acid, 6-oxo-2,3-diphenyl-, 1,1-dimethylethyl ester, (2S,3R)-

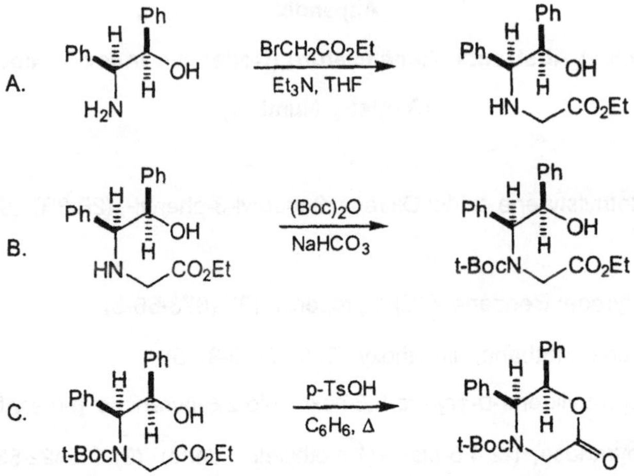

A. $BrCH_2CO_2Et$, Et_3N, THF

B. $(Boc)_2O$, $NaHCO_3$

C. p-TsOH, C_6H_6, Δ

Submitted by Robert M. Williams, Peter J. Sinclair, Duane E. DeMong, Daimo Chen and Dongguan Zhai[1]

Checked by Wenlin Lee and Marvin J. Miller

1. Procedure

A. *Ethyl (1'S,2'R)-N-(1',2'-diphenyl-2'-hydroxyethyl)glycinate.* A 1-L, three-necked, round-bottomed flask, equipped with a magnetic stirring bar, pressure-equalizing addition funnel, and two stoppers, is charged with 25 g (0.117 mol) of (1S,2R)-1,2-diphenyl-2-hydroxyethylamine (Note 1), 29 g (0.176 mol) of ethyl bromoacetate and 455 mL of anhydrous tetrahydrofuran (THF) (Note 2). To the stirred mixture, 24 g (0.234 mol) of triethylamine (Note 3) is added dropwise via the addition funnel. After stirring vigorously for 20 hr at room temperature, the solids are removed by filtration and washed three times with THF (Note 4). The clear yellowish filtrate and washes are concentrated on a rotary evaporator (to remove solvent, excess triethylamine, and ethyl bromoacetate) to afford an off-white solid, which is collected with a Buchner funnel, then washed with water (Note 5) to dissolve residual salts. The residue is air-dried in the filter funnel with water aspirator vacuum. The crude product is dissolved in 117 mL of boiling absolute ethanol (EtOH) and stored at room temperature overnight. White crystals are collected by filtration, washed with ice-cold (0°C) absolute EtOH carefully, and dried at 50°C (Note 6). The above described procedure is done in duplicate, providing ethyl (1'S,2'R)-N-(1',2'-diphenyl-2'-hydroxyethyl)glycinate in yields of 29.01 g and 28.96 g, (average yield: 83%) (Notes 7, 8).

B. *Ethyl (1'S,2'R)-N-tert-butyloxycarbonyl-N-(1',2'-diphenyl-2'-hydroxyethyl-glycinate.*[3] A 1-L, three-necked, round-bottomed flask, equipped with a magnetic stirring bar, reflux condenser, and two stoppers, is charged with 167 mL of a solution of saturated aqueous sodium bicarbonate ($NaHCO_3$), 34.2 g (0.585 mol) of sodium chloride (NaCl), 350 mL of chloroform ($CHCl_3$) (Note 9), 25 g (0.084 mol) of ethyl (1'S,2'R)-N-(1',2'-diphenyl-2'-hydroxyethyl)glycinate, and 36.5 g (0.167 mol) of di-tert-butyl dicarbonate (Note 10). The resulting mixture is then heated to reflux for 20 hr and vigorously stirred.

The reaction mixture is placed in a separatory funnel; the aqueous phase is separated and extracted twice with chloroform ($CHCl_3$). The organic phases are combined, washed twice with water, and dried with anhydrous sodium sulfate overnight. After filtration to remove sodium sulfate, concentration under vacuum affords a residue, which is heated (5 mm, 130°C oil bath) to remove and recover excess di-tert-butyl dicarbonate (Note 11). A total of 33.2 g (83 mmol) of crude ethyl (1'S,2'R)-N-tert-butyloxycarbonyl-N-(1',2'-diphenyl-2'-hydroxyethyl)glycinate, is obtained as an oil and used directly in the subsequent lactonization reaction.

C. *(5S,6R)-4-tert-Butoxycarbonyl-5,6-diphenylmorpholin-2-one.* A 2-L, round-bottomed flask, equipped with a magnetic stirring bar, is charged with 33.2 g (0.083 mol) of crude ethyl (1'S,2'R)-N-tert-butyloxycarbonyl-N-(1',2'-diphenyl-2'-hydroxyethyl) glycinate, 1.90 g (0.010 mol) of p-toluenesulfonic acid monohydrate (Note 12), and 670 mL of benzene (Notes 13, 14). The reaction flask is fitted with a Soxhlet extractor, which is packed with 65 g of calcium chloride ($CaCl_2$, anhydrous, 30 mesh) and a condenser with a drying tube. The reaction mixture is heated to reflux with stirring for 10 hr (Note 15). After the first 4 hr, the $CaCl_2$ in the Soxhlet extractor is replaced; the total amount of $CaCl_2$ used in the reaction is 130 g. Ultimately, the reaction mixture becomes a suspension. Removal of benzene under reduced pressure (Note 16) yields white solids that are dissolved in 310 mL of dichloromethane (CH_2Cl_2). The solution is washed with water and 5% aqueous $NaHCO_3$ in a separatory funnel until the CH_2Cl_2 solution becomes clear. After removal of CH_2Cl_2 under vacuum, the resulting white crystals are dissolved by boiling in 600 mL of absolute EtOH; the resulting solution is allowed to stand at room temperature overnight. The white crystals are collected by filtration, washed with ice-cold EtOH, and dried at 60°C. Concentration of the mother liquor by rotary evaporation affords a residue that can also be recrystallized in absolute EtOH to provide additional product. The above described procedure is performed in duplicate, providing (5S,6R)-4-tert-butyloxycarbonyl-5,6-diphenylmorpholin-2-one in yields of 21.7 g and 20.3 g (average yield for the two steps: 71%) (Notes 17-19).

2. Notes

1. (1S,2R)- and (1R,2S)-1,2-Diphenyl-2-hydroxyethylamine (>98% ee) were prepared according to the procedure of Tishler, et al.[2] The checkers obtained the amino alcohols from Aldrich Chemical Co., Inc.

2. Tetrahydrofuran was dried by distillation from sodium-benzophenone ketyl.

3. Triethylamine was dried by distillation from CaH_2.

4. Reagent grade or technical grade (i.e. undistilled) THF can be used for this wash.

5. Extensive washing in a Buchner funnel with water is recommended.

6. The product was dried in a crystallizing dish in a vacuum oven.

7. The physical and spectroscopic properties for ethyl (1'S,2'R)-N-(1',2'-hydroxyethyl)glycinate are as follows: mp. 127-128°C, $[\alpha]_D^{25}$ +24.4 (*c* 5.5, CH_2Cl_2); IR (NaCl, $CDCl_3$) cm^{-1}: 3840-3430, 3330, 3080, 3045, 2995, 2940, 1750, 1460, 1385, 1210, 1035, 915, 740; ^1H NMR (270 MHz, $CDCl_3$,) δ: 1.20 (3H, t, J = 7.1), 2.2 (2H, br s), 3.15 (1H, 1/2 AB q, J = 17.5), 3.29 (1H, 1/2 AB q, J = 17.5), 3.95 (1H, d, J = 6.0), 4.11 (2H, q, J = 7.1), 4.80 (1H, d, J = 6.0), 7.17-7.32 (10H, m); Anal. Calcd. for $C_{18}H_{21}NO_3$: C, 72.22; H, 7.07; N, 4.68. Found: C, 72.31; H, 7.16; N, 4.56.

8. From the antipode (1R,2S)-1,2-diphenyl-2-hydroxyethylamine is obtained the corresponding enantiomer ethyl (1'R,2'S)-N-(1',2'-diphenyl-2'-hydroxyethyl)-glycinate. mp 127-128°C, $[\alpha]_D^{25}$ –24.4 (*c* 5.5, CH_2Cl_2). This series, also performed in duplicate, provides yields of 29.2 g and 29.5 g (average yield: 84%).

9. Technical grade chloroform was used.

10. Di-tert-butyl dicarbonate was purchased from Aldrich Chemical Co., Inc.

11. The recovered di-tert-butyl dicarbonate is very pure and can be reused.

12. p-Toluenesulfonic acid monohydrate was purchased from Aldrich Chemical Co., Inc.

13. ACS grade benzene was purchased from Fisher Scientific.

14. The checkers found that 770 mL of cyclohexane could also be used as the solvent for this step, avoiding the use of large amounts of toxic benzene. The crude Step B product, a white solid, very slowly dissolves in refluxing cyclohexane; upon conversion to the morpholinone, which is insoluble in cyclohexane, the product precipitates out of solution. The reaction mixture remains a suspension throughout the reaction, but the texture of the solids shows a subtle change over time, indicating the progress of the reaction. The color of the solution also turns slightly yellow. The reaction time is extended to 17 hr and the $CaCl_2$ pellets in the Soxhlet extractor are changed after the first 5-6 hr. Work up of the reaction is identical to that described in the procedure.

15. The Soxhlet extractor does not require the standard Soxhlet filter thimbles; cotton above and below the $CaCl_2$ pellets is sufficient.

16. *Caution!* Care should be exercised in performing the reaction and concentrating benzene in a fume hood.

17. The physical and spectroscopic properties for (5S,6R)-4-tert-butyloxycarbonyl-5,6-diphenylmorpholin-2-one are as follows: mp 207°C; $[\alpha]_D^{25}$ +86.3 (*c* 5.6, CH_2Cl_2); IR (NaCl, CH_2Cl_2) cm^{-1}: 3050, 2975. 1755, 1690, 1380, 1255, 1150, 1100, 1045; ^1H NMR (200 MHz) (DMSO-d_6 vs. DMSO) (120°C) δ: 1.25 (9H, s), 4.52 (2H, d, J = 1.1), 5.16 (1H, d, J = 3.0), 6.17 (1H, d, J = 3.0), 6.63-6.68 (2H, m), 7.0-7.3 (8H, m); Anal. calcd. for $C_{21}H_{23}NO_4$: C, 71.37; H, 6.56; N, 3.96. Found: C, 71.08; H, 6.44; N, 3.98. The checkers obtained 22.7 g, (77%) after the first recrystallization; mp 205-207°C; $[\alpha]_D^{25}$ −86.0 (*c* 1.0, CH_2Cl_2).

18. From the antipode, ethyl (1'R,2'S)-N-(1',2'-diphenyl-2'-hydroxyethyl)-glycinate, is obtained the corresponding enantiomer, (5R,6S)-4-tert-butyloxycarbonyl-5,6-diphenylmorpholin-2-one, mp 207°C, $[\alpha]_D^{25}$ +85.8 (*c* 5.7, CH_2Cl_2). This series, also performed in duplicate, provides yields of 24.0 g and 23.6 g, respectively (average yield for the two steps: 80%).

19. Optical rotations were measured on a Rudolf Research Autopol III automatic polarimeter operating at 589 nm, corresponding to the sodium D line.

Waste Disposal Information

All toxic materials were disposed of in accordance with "Prudent Practices in the Laboratory"; National Academy Press; Washington, DC, 1995.

3. Discussion

(+)- and (-)-4-tert-Butoxycarbonyl-5,6-diphenylmorpholin-2-one can be used for the asymmetric synthesis of α-amino acids.[3] The methylene carbon of the morpholinone can be efficiently brominated with NBS in CCl_4 to give a single diastereomeric bromide (anti). This bromo lactone serves as a useful electrophilic glycine synthon that undergoes stereoselective coupling with a variety of organometallic reagents. In general, coupling proceeds with net retention of stereochemistry furnishing the crystalline anti-homologation products. Dissolving metal reduction (lithium or sodium in liquid NH_3/EtOH/THF) directly furnishes the corresponding N-tert-butoxycarbonyl-protected amino acid derivative. Alternatively, the N-t-Boc group can be removed (TFA or HCO_2H) and catalytic hydrogenation ($PdCl_2$/EtOH/THF, 25°C, 40 psi) furnishes the free zwitterionic amino acid. In some cases, coupling has been observed to proceed with net inversion of stereochemistry to furnish syn-homologation products as oils. In cases where modest stereoselectivity is observed, the syn- and anti-isomers can be readily separated by chromatography. If the coupling is anti-selective (which is typical), a simple recrystallization will consistently give the final amino acid in >98% ee.

The lactones are quite stable to storage and handling (shelf life at room temperature >16 years). Treatment of these substances with either lithium, sodium or potassium hexamethyldisilazane in THF at –78°C generates the corresponding enolate that can be stereoselectively alkylated with an appropriate electrophile to give anti-homologated products. Similar reductive processing of these materials provides amino acid derivatives in high enantiomeric excess.

The present method offers a versatile and practical protocol for preparing either D- or L- amino acids and offers the unique advantage of allowing direct access to the N-t-Boc derivatives. Conversion of the diphenyl amino alcohol moiety into the desired product and the water-insoluble bibenzyl, which is removed by simple extraction into ether or pentane, by either reductive protocol (1 and 2, shown below) is noteworthy. The water-soluble zwitterions, or in the case of the N-t-Boc amino acids, the N-t-Boc carboxylate salts, remain in the aqueous phase and are conveniently isolated in high chemical purity without, in most cases, the need for additional time-consuming and expensive chromatographic purification steps. With the oxidative methods (3 and 4), the diphenyl amino alcohol moiety is converted into two equivalents of benzaldehyde which can generally be separated from the amino acid by extraction. The ease and convenience of processing the derivatized morpolinones to the desired amino acids is testament to the flexibility and reliability of this methodology.

Method	Conditions	Type of 'R' Group	Amino Acid
1) Birch Reduction	(Li° or Na°, $NH_{3(l)}$)	Unsaturated; Saturated, Aliphatic	N-t-Boc
2) Catalytic hydrogenation (after removal of the Boc group)	(H_2 / Pd-C, EtOH)	Saturated, Aliphatic	Zwitterion
3) Oxidation	1) H^+; 2) $NaIO_4$	Aromatic; Saturated, Aliphatic	Zwitterion
4) Oxidation	1) H^+; 2) $Pb(OAc)_4$	Aromatic β-Arylcyclopropyl Saturated, Aliphatic	Methyl ester

As illustrated in the accompanying procedure, (N-t-Boc)allylglycine, this morpholinone has proven to be one of the most versatile and useful chiral glycine equivalents from which a large variety of amino acids can be prepared. Several other, recent asymmetric electrophilic glycine enolate and cation equivalents have been devised.[4]

Glycine synthons

glycine electrophile | glycine enolate | radical | [1,3]-dipole | phosphonate | α,β-dehydro- | acetal → oxonium

α-amino acids

peptide isosteres

As shown in the scheme above, no other chiral glycine equivalent (from the laboratories of Seebach, Oppolzer, Evans, Myers, Schöllkopf, etc.) can provide the range of reactivity of this substance.

The chemistry of these compounds was reviewed in 1992[3j] and again in 1995[3q] and there are numerous publications and patent citations in the literature from other laboratories that have employed these synthons; a few examples are cited.[5–17] Finally, the N-t-Boc-protected morpholinones have been used extensively in our laboratory to prepare a variety of amino acids and peptide isosteres as shown below.[3] These examples serve to illustrate the great versatility that these glycine derivatives offer for preparing structurally diverse amino acids in high optical purity.

spirotryprostatin

2,7-diaminosuberic acid

meso-2,6-diaminopimelic acid

coronamic acid

β,γ-Unsaturated-α-amino acids

References

1. Department of Chemistry, Colorado State University, Fort Collins, Colorado 80523. This material is based upon work supported by the National Science Foundation under Grant No. 0202827.

2. Weijlard, J.; Pfister, K.; Swanezy, E. F.; Robinson, C. A.; Tishler, M. *J. Am. Chem. Soc.* **1951**, *73*, 1216.

3. (a) Sinclair, P. J.; Zhai, D.; Reibenspies, J.; Williams, R. M. *J. Am. Chem. Soc.* **1986**, *108*, 1103; (b) Williams, R. M.; Zhai, D.; Sinclair, P. J. *J. Org. Chem.* **1986**, *51*, 5021; (c) Williams, R. M.; Sinclair, P. J.; Zhai, W. *J. Am. Chem. Soc.* **1988**, *110*, 482; (d) Williams, R. M.; Im, M.-N. *Tetrahedron Lett.* **1988**, *29*, 6075 (e) Williams, R. M.; Sinclair, P. J.; Zhai, D.; Chen, D. *J. Am. Chem. Soc.* **1988**, *110*, 1547; (e) Williams, R. M.; Zhai, W. *Tetrahedron* **1988**, *44*, 5425; (f) Williams, R. M.; Hendrix, J. A. *J. Org. Chem.* **1990**, *55*, 3723; (g) Williams, R. M.; Im, M.-N.; Cao, J. *J. Am. Chem. Soc.* **1991**, *113*, 6976; (h) Williams, R. M.; Fegley, G. J. *J. Am. Chem. Soc.* **1991**, *113*, 8796; (i) Williams, R. M.; Im, M.-N. *J. Am. Chem. Soc.* **1991**, *113*, 9276; (j) Williams, R. M. *Aldrichimica Acta* **1992**, *25*, 11; (k) Williams, R. M.; Yuan, C. *J. Org. Chem.* **1992**, *57*, 6519; (l) Williams, R. M.; Fegley, G. J. *Tetrahedron Lett.* **1992**, *33*, 6755; (m) Williams, R. M.; Zhai, W.; Aldous, D. J.; Aldous, S. C. *J. Org. Chem.* **1992**, *57*, 6527; (n) Williams, R. M.; Fegley, G. J. *J. Org. Chem.* **1993**, *58*, 6933; (o) Williams, R. M.; Yuan, C. *J. Org. Chem.* **1994**, *59*, 6190; (p) Williams, R. M.; Colson, P. J.; Zhai, W. *Tetrahedron Lett.* **1994**, *35*, 9371; (q) Williams, R. M., *Advances in Asymmetric Synthesis*, JAI Press, A. Hassner, Ed. **1995**, Vol. 1, p. 45; (r) Williams, R. M.; Fegley, G. J.; Gallegos, R.; Schaeffer, F.; Pruess, D. L. *Tetrahedron* **1996**, *52*, 1149; (s) Bender, D. M.; Williams, R. M. *J. Org. Chem.* **1997**, *62*, 6690; (t) Williams, R. M.; Liu, J. *J. Org. Chem.* **1998**, *63*, 2130; (u) Williams, R. M.; Aoyagi, Y. *Tetrahedron* **1998**, *54*, 10419; (v) Scott, J. D.; Tippie, T. N.; Williams, R. M. *Tetrahedron Lett.* **1998**, *39*,

3659; (w) Aoyagi, Y.; Williams, R.M. *Synlett* **1998**, 1099; (x) Aoyagi, Y.; Williams, R. M. *Tetrahedron* **1998**, *54*, 13045; (y) Williams, R. M, *Peptidomimetics Protocols,* Kazmierski, W., Ed., *Methods in Molecular Medicine* **1999**, Vol. 23, Chapter 19, p. 339, Humana Press; (z) Sebahar, P. R.; Williams, R. M. *J. Am. Chem. Soc.* **2000**, *122*, 5666.

4. For reviews, see: (a) *Synthesis of Optically Active α-Amino Acids*, Robert M. Williams, (Organic Chemistry Series, J. E. Baldwin, Series Editor); **1989**, p. 410, Pergamon Press, New York; (b) Duthaler, R. O. *Tetrahedron* **1994**, *50*, 1539; (c) O'Donnell, M. J., Ed., *Tetrahedron Symposia in Print* Tetrahedron **1988**, *44*, 5253; (d) Cativiela, C.; Diaz-de-Villegas, M. D. *Tetrahedron: Asymm.*.**1998**, *9*, 3517; (e) Cativiela, C.; Diaz-de-Villegas, M. D. *Tetrahedron: Asymm.* **2000**, *11*, 645.
5. Ramer, S. E.; Cheng, H.; Vederas, J. C. *Pure Appl. Chem.* **1989**, *61*, 489.
6. Reno, D. S.; Lotz, B. T.; Miller, M. J. *Tetrahedron Lett.* **1990**, *31*, 827.
7. Dudycz, L. W. *Nucleosides and Nucleotides*, **1991**, *10*, 329.
8. Dong, Z. *Tetrahedron Lett.* **1992**, *33*, 7725.
9. Baldwin, J. E.; Lee, V.; Schofield, C. J. *Heterocycles* **1992**, *34*, 903.
10. Baldwin, J. E.; Lee, V.; Schofield, C. J. *Synlett* **1992**, 249.
11. Schow, S. R.; DeJoy, S. Q.; Wick, M. M.; Kerwar, S. S. *J. Org. Chem.* **1994**, *59*, 6850.
12. Maggini, M.; Scorrano, G.; Bianco, A.; Toniolo, C.; Sijbesma, R. P.; Wudl, F.; Prato, M. *J. Chem. Soc. Chem. Comm.* **1994**, 305.
13. Schuerman, M. A.; Keverline, K. I.; Hiskey, R. G. *Tetrahedron Lett.* **1995**, *36*, 825.
14. Solas, D.; Hale, R. L.; Patel, D. V. *J. Org. Chem.* **1996**, *61*, 1537.
15. Hutton, C. A.; White, J. M. *Tetrahedron Lett.* **1997**, *38*, 1643.
16. Lee, S-H.; Nam, S-W. *Bull. Korean Chem. Soc.* **1998**, *19*, 613.
17. van den Nieuwendijk, A. M. C. H.; Benningshof, J. C. J.; Wegmann, V.; Bank, R. A.; te Koppele, J. M.; Brussee, J.; van der Gen, A. *Bioorg. Med. Chem. Lett.* **1999**, *9*, 1673.

Appendix

Chemical Abstracts Nomenclature (Collective Index Number); (Registry Number)

Ethyl bromoacetate: Acetic acid, bromo-, ethyl ester (8,9); (105-36-2)

Triethylamine: N,N-Diethylethanamine (9); (121-44-8)

Di-tert-butyl dicarbonate: Dicarbonic acid, bis(1,1-dimethylethyl) ester (9); (24424-99-5)

p-Toluenesulfonic acid: Benzenesulfonic acid, 4-methyl-, monohydrate (9); (6192-52-5)

(1S,2R)-1,2-Diphenyl-2-hydroxyethylamine: Benzeneethanol, β-amino-α-phenyl-, (αR,βS)-rel- (9); (23412-95-5)

(1R,2S)-1,2-Diphenyl-2-hydroxyethylamine: Benzeneethanol, β-amino-α-phenyl-, (αS,βR)- (9); (23364-44-5)

Ethyl (1'S,2'R)-N-(1',2'-diphenyl-2'-hydroxyethyl)glycinate: Glycine, N-(2-hydroxy-1,2-diphenylethyl)-, ethyl ester, [R-(R*,S*)]- (9); (100678-82-8)

Ethyl (1'R,2'S)-N-(1',2'-diphenyl-2'-hydroxyethyl)glycinate: Glycine, N-(2-hydroxy-1,2-diphenylethyl)-, ethyl ester, [S-(R*,S*)]- (9); (112835-62-8)

(1'S,2'R)-Ethyl N-tert-butyloxycarbonyl-N-(1',2'-diphenyl-2'-hydroxyethyl)glycinate: Glycine, N-[(1,1-dimethylethoxy)carbonyl]-N-(2-hydroxy-1,2-diphenylethyl)-, ethyl ester, [R-(R*,S*)]- (9); (112741-70-5)

(1'R,2'S)-Ethyl N-tert-butyloxycarbonyl-N-(1',2'-diphenyl-2'-hydroxyethyl)glycinate: Glycine, N-[(1,1-dimethylethoxy)carbonyl]-N-(2-hydroxy-1,2-diphenylethyl)-, ethyl ester, [S-(R*,S*)]- (9); (112741-73-8)

(5S,6R)-4-tert-Butyloxycarbonyl-5,6-diphenylmorpholin-2-one: 4-Morpholinecarboxylic acid, 6-oxo-2,3-diphenyl-, 1,1-dimethylethyl ester, (2R-cis)- (9); (173397-90-5)

(5R,6S)-4-tert-Butyloxycarbonyl-5,6-diphenylmorpholin-2-one: 4-Morpholinecarboxylic acid, 6-oxo-2,3-diphenyl-, 1,1-dimethylethyl ester, (2S-cis)- (9); (112741-50-1)

EFFICIENT ASYMMETRIC SYNTHESIS OF N-tert-BUTOXYCARBONYL α-AMINOACIDS USING 4-tert-BUTOXYCARBONYL-5,6-DIPHENYLMORPHOLIN-2-ONE: (R)-(N-tert-BUTOXYCARBONYL)ALLYLGLYCINE

(4-Pentenoic acid, 2-[[(1,1-dimethylethoxy)carbonyl]amino]-, (2R)-)

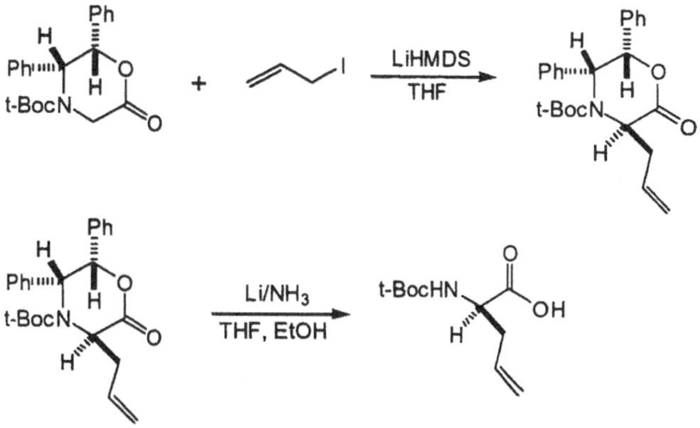

Submitted by Robert M. Williams,[1] Peter J. Sinclair, and Duane E. DeMong.
Checked by Wenlin Lee and Marvin J. Miller

1. Procedure

A. *(3R,5R,6S)-4-tert-Butoxycarbonyl-5,6-diphenyl-3-(1'-prop-2'-enyl)morpholin-2-one.* A 250-mL, single neck, round-bottomed flask equipped with a magnetic stirring bar is charged with 1.0 g (2.83 mmol) of (5R,6S)-4-tert-butoxycarbonyl-5,6-diphenylmorpholin-2-one and placed under vacuum for 3 hr. The evacuated flask is purged with argon, a rubber septum with an argon inlet is attached to the neck, and tetrahydrofuran (THF) (50 mL) is added via syringe (Note 1). After dissolution of the solid by stirring, 2 mL (10 mmol) of allyl iodide (Note 2) is added via syringe. With stirring, the flask and its contents are cooled to –78°C with a dry ice/acetone bath, then 2.8 mL (2.83 mmol) of lithium bis(trimethylsilyl)amide (1M in THF) is added dropwise (Note 3). The mixture is stirred for 1 hr at –78°C (Note 4), then the solution is quenched by the addition of 100 mL of water. The mixture is extracted with 250 mL of ethyl acetate in a separatory funnel. After separation of the aqueous phase, the organic phase is washed with brine, dried over anhydrous magnesium sulfate ($MgSO_4$), and concentrated by rotary evaporation under vacuum to afford an orange oil. The crude product is dissolved in a minimum amount of 4:1 hexanes:ethyl acetate, and purified by flash chromatography using 70 g of 35-75 mesh silica gel (eluting with 4:1 hexane:ethyl acetate followed by 2:1 hexanes:ethyl acetate). Removal of solvent from the appropriate fractions by rotary evaporation at room temperature affords 997 mg (90%) of (3R,5R,6S)-4-tert-butoxycarbonyl-5,6-diphenyl-3-(1'-prop-2'-enyl)-morpholin-2-one as a white solid (Notes 5, 6).

B. *(R)-(N-tert-Butoxycarbonyl)allylglycine*. A flame-dried, 100-mL, two-necked, round-bottomed flask equipped with a magnetic stirring bar is charged with 35 mL of liquid ammonia and 124 mg (17.8 mmol) of lithium metal (Note 7). The resulting solution is stirred at -33°C. Meanwhile, a flame-dried 25-mL, round-bottomed flask fitted with a rubber stopper and an argon inlet is charged with 539 mg (1.37 mmol) of (3R,5R,6S)-4-tert-butoxycarbonyl-5,6-diphenyl-3-(1'-prop-2'enyl)morpholin-2-one (p.18), 0.150 mL of absolute ethanol (Note 8), and 5 mL of anhydrous THF (Note 1). The resulting mixture is then transferred via cannula to the lithium/liquid ammonia solution. After 20 min, the reaction is carefully quenched with solid ammonium chloride and allowed to warm to room temperature (Note 9). After evaporation of the remaining ammonia, the residue is diluted with 35 mL of water and extracted twice with 35-mL portions of diethyl ether (Note 10). The aqueous phase is carefully acidified with 1N HCl to a pH of 2 while stirring with 35 mL of ethyl acetate (Note 11). Following separation of the organic phase, the aqueous phase is extracted three times with 20-mL portions of ethyl acetate. The organic extracts are combined, dried over $MgSO_4$, filtered, and concentrated under vacuum to yield 178 mg (60%) of crude (R)-(N-tert-butoxycarbonyl)allylglycine (% ee ≥ 96%) as a viscous oil (Notes 12-15).

2. Notes

1. Anhydrous tetrahydrofuran is obtained by distilling from sodium benzophenone ketyl.

2. Allyl iodide (98%) can be purchased from Aldrich Chemical Co., Inc.

3. Lithium bis(trimethylsilyl)amide (1M solution in tetrahydrofuran) can be purchased from Aldrich Chemical Co., Inc.

4. Phosphomolybdic acid (PMA) dissolved in ethanol is a suitable reagent for visualizing the product on TLC. After dipping the TLC plate in the PMA solution, the chromatogram is developed by heating on a hot plate.

5. Trace amounts of impurities can be removed by recrystallizing the product from either diethyl ether/hexanes or dichloromethane/hexanes. The yield obtained by the checker after the first recrystallization was 65%, although an additional crop of the product could be obtained by recrystallization of the mother liquor.

6. Analytical and spectral data are as follows: mp. 177-178°C, $[\alpha]_D^{25}$ +45.7 (c 1.34, CH_2Cl_2); ^1H NMR (300 MHz) (DMSO-d_6 vs. DMSO) (120°C) δ: 1.22 (9H, s broad); 2.87 (2H, m); 4.90 (1H, dd, J = 7.3, 7.0); 5.17-5.29 (3H, m); 5.89-6.04 (1H, m); 6.20 (1H, d, J = 3.3); 6.59-6.62 (2H, m); 7.1-7.3 (8H, m). IR (NaCl, CH_2Cl_2) cm^{-1}: 3050, 2970, 2920, 1755, 1690, 1375, 1350, 1260, 1155, 1110. Analysis (recrystallized from diethyl ether/hexanes or dichloromethane/hexanes); calcd for $C_{28}H_{27}NO_4$: C, 73.26; H, 6.92; N, 3.56. Found: C, 72.37; H, 6.85; N, 3.68. Note that the NMR spectrum must be recorded at

high temperature due to slow conformational exchange of the urethane group on the NMR time scale.

7. Ammonia is distilled from sodium metal immediately prior to use.

8. Absolute ethanol is used as received from Pharmco Products Incorporated.

9. When quenching the reaction and warming to room temperature, caution must be exercised in order to prevent the reaction mixture from overflowing the flask if ammonium chloride is added too quickly or the reaction is warmed too rapidly.

10. Reagent grade diethyl ether is used as received from Fisher Scientific.

11. An Orion Research model 301 analog pH meter was used.

12. The crude product is typically analytically pure; however, the product can be further purified by flash chromatography using 15 g of 35-75 mesh silica gel (elution with 60 mL of 5% methanol in dichloromethane).

13. Analytical and spectral data are as follows: $[\alpha]_D^{25}$ = +3.3 (*c* ,1.5, CH_2Cl_2); ^1H NMR (300 MHz) (DMSO-d_6) (60°C) δ TMS: 1.39 (9H, s); 2.29-2.46 (2H, m); 3.92-4.0 (1H, m); 5.02-5.13 (2H, m); 5.77 (1H, dddd, J = 17.2, 10.3, 7.0, 7.0); 6.73 (1H, br s); 12.13 (1H, br s). IR (neat) cm^{-1}: 3430, 3050, 2980, 1715, 1500, 1370, 1265, 1155; mass spectrum (NH_3, CI) *m/e* 232.9 (M^+ +18, 1.8), 215.9 (M^+ +1, 2.1), 214.9 (M^+, 0.3), 116.0 (62.9).

14. Optical rotations were measured on a Rudolf Research Autopol III automatic polarimeter operating at 589 nm, corresponding to the sodium D line.

15. The enantiomeric excess of the product is determined by Mosher amide formation and NMR analysis as follows: (R)-(N-tert-butoxycarbonyl)allylglycine is dissolved in 3 mL of 1M ethanolic HCl and heated to reflux for 2 hr. The reaction is cooled and the solvent removed under vacuum. The crude amino ester hydrochloride is dissolved in 0.3 mL of pyridine and 0.3 mL of CCl_4. To this mixture is added 25 mg (0.10 mmol) of (+)-α-methoxy-α-(trifluoromethyl)phenylacetyl chloride and the reaction is stirred for 6 hr at room temperature. The reaction is quenched with 1 mL water, taken up in diethyl ether, and washed consecutively with 1M HCl, saturated aqueous $NaHCO_3$, and water. The ether

layer is dried over anhydrous $MgSO_4$, filtered and evaporated. The Mosher amide formation is repeated with (±)-α-methoxy-α-(trifluoromethyl)phenylacetyl chloride, and, now with this reference standard, the two products are compared by ^1H NMR.

Waste Disposal Information

All toxic materials were disposed of in accordance with "Prudent Practices in the Laboratory"; National Academy Press; Washington, DC, 1995.

3. Discussion

The morpholinone glycine synthon, which can be obtained in both enantiomeric forms, can be employed as a versatile scaffold for the preparation of a wide range of N-t-Boc protected R- and S-α-amino acids.[2-4] Allylglycine, the example selected here to demonstrate this methodology, is useful for the construction of several amino acid derivatives, such as 2,7-diaminosuberic acid[5,6] and mechanism-based inhibitors of pyridoxal-dependent enzymes.[7] Allylglycine has also served as a starting material for the synthesis of several natural products.[8-11] The current procedure is the only direct method of synthesizing this N-t-Boc-protected amino acid. Other reports have appeared for preparing this amino acid[12,13] including an Organic Syntheses procedure by Myers.[14] The use of the dissolving metal reduction to remove the chiral auxiliary allows the direct isolation of the N-t-Boc-protected amino acid, thus eliminating the need to protect the free amino acid in a subsequent step. The by-product of the reduction of the diphenylamino alcohol moiety is bibenzyl, which is insoluble in water and conveniently removed by extraction as described above. The current procedure enables these alkylations to be carried out on a preparative scale in good yield and excellent ee.

References

1. Department of Chemistry, Colorado State University, Fort Collins, Colorado 80523. This material is based upon work supported by the National Science Foundation under Grant No. 0202827.
2. Williams, R. M., *Aldrichimica Acta* **1992**, *25*, 11.
3. Williams, R. M., *Advances in Asymmetric Synthesis*, A. Hassner, Ed., JAI Press **1995**, Vol. 1, p 45.
4. Williams, R. M, *Peptidomimetics Protocols,* Kazmierski, W., Ed., *Methods in Molecular Medicine* **1999,** Vol. 23, Chapter 19, p. 339, Humana Press.
5. Williams, R. M.; Liu, J. *J. Org. Chem.* **1998**, *63*, 2130.
6. Gao, Y.; Lane-Bell, P.; Vederas, J. C. *J. Org. Chem.* **1998**, *63*, 2133.
7. Johnston, M.; Raines, R.; Walsh, C.; Firestone, R. A. *J. Am. Chem. Soc.* **1980**, *102*, 4241.
8. Hutton, C. A.; White, J.M. *Tetrahedron Lett.* **1997**, *38*, 1643.
9. Kurokawa, N.; Ohfune, Y. *Tetrahedron* **1993**, *49*, 6195.
10. Baldwin, J. E.; Bradley, M.; Turner, N. J.; Adlington, R.M.; Pitt, A. R.; Sheridan, H. *Tetrahedron* **1991**, *47*, 8203.
11. Schneider, H.; Sigmund, G.; Schricker, B.; Thirring, K.; Berner, H. *J. Org. Chem.* **1993**, *58*, 683.
12. Hamon, D. P. G.; Massy-Westropp, R. A.; Razzino, P. *Tetrahedron* **1995**, *51*, 4183.
13. R. M. Williams, *Synthesis of Optically Active α-Amino Acids* (Organic Chemistry Series, J. E. Baldwin, Series Editor) **1989**, p. 410, Pergamon Press, New York.
14. Myers, A. G.; Gleason, J.L. *Org. Synth.* **1999**, *76*, 57.

Appendix

Chemical Abstracts Nomenclature (Collective Index Number);

(Registry Number)

(R)-(N-tert-Butoxycarbonyl)allylglycine: 4-Pentenoic acid,

2-[[(1,1-dimethylethoxy)carbonyl]amino]-, (2R)- (9); (170899-08-8)

(5R, 6S)-4-tert-Butyloxycarbonyl-5,6-diphenylmorpholin-2-one:

4-Morpholinecarboxylic acid, 2-oxo-5,6-diphenyl-, 1,1-dimethylethyl ester, (2S-cis)- (9);

(112741-50-1)

Allyl iodide: 1-Propene, 3-iodo- (9); (556-56-9)

(3R,5R,6S)-4-tert-Butoxycarbonyl-5,6-diphenyl-3-(1'-prop-2'-enyl)morpholin-2-one:

4-Morpholinecarboxylic acid, 2-oxo-5,6-diphenyl-3-(2-propenyl)-, 1,1-dimethylethyl ester,

3R-(3α,5β,6β)]- (9); (143140-32-3)

Lithium bis(trimethylsilyl)amide: Silanamine, 1,1,1-trimethyl-N-(trimethylsilyl)-, lithium salt

(9); (4039-32-1)

Lithium: Lithium (8,9); (7439-93-2)

PREPARATION OF O-ALLYL-N-(9-ANTHRACENYLMETHYL)CINCHONIDINIUM BROMIDE AS A PHASE TRANSFER CATALYST FOR THE ENANTIOSELECTIVE ALKYLATION OF GLYCINE BENZOPHENONE IMINE tert-BUTYL ESTER: (4S)-2-(BENZHYDRYLIDENAMINO)PENTANEDIOIC ACID, 1-tert-BUTYL ESTER-5-METHYL ESTER

[Cinchonanium, 1-(9-anthracenylmethyl)-9-(2-propenyloxy)-, bromide, (8α,9R)- and L-Glutamic acid, N-(diphenylmethylene)-, 1-(1,1-dimethylethyl) 5-methyl ester]

A. Toluene, reflux

B. Allyl bromide, 50% aq. NaOH, CH_2Cl_2, r.t.

C. Alkaloid Catalyst, Methyl acrylate, $CsOH \cdot H_2O$, CH_2Cl_2, -78°C, 4 hr

Submitted by E. J. Corey[1] and Mark C. Noe[2]

Modified and checked by S. Springfield, J. Amato, and E. J. J. Grabowski

1. Procedure

A. *N-(9-Anthracenylmethyl)cinchonidinium chloride* (Note 1). A 2-L, three-necked flask, equipped with an overhead stirrer, reflux condenser and nitrogen inlet, is charged with 78.2 g (266 mmol) of cinchonidine (Note 2), 63.0 g (278 mmol) of 9-chloromethylanthracene (Note 3), and 800 mL of toluene (Note 4). The mixture is stirred and heated to reflux employing a bath temperature of ~130°C for 3 hr. Within 30 min all of the material dissolves, then the product begins to crystallize. At the end of the reflux period the mixture is cooled to ~30°C and 800 ml of diethyl ether is added over 10 min. The slurry is stirred for 20 min, filtered through a coarse sintered glass, fritted funnel, washed with 250 ml of diethyl ether and dried at 50°C under vacuum overnight to afford 144.4 g (235 mmol, 88%) of N-(9-anthracenylmethyl)cinchonidinium chloride as a crystalline, light yellow solid (Notes 1, 5), which is sufficiently pure for use in the next step.

B. *O-Allyl-N-(9-Anthracenylmethyl)cinchonidinium bromide*. A 1-L, three-necked flask, equipped with an overhead stirrer and nitrogen inlet, is charged with 49.5 g (80.7 mmol) of N-(9-anthracenylmethyl)cinchonidinium chloride–toluene solvate, 400 mL of methylene chloride (CH_2Cl_2), 25 mL of allyl bromide (35.0 g, 289 mmol) (Note 6) and 50 mL of 50% aqueous potassium hydroxide (Note 7). The mixture is stirred vigorously for 4 hr, then diluted with 400 mL of water and stirred for 5 min. After separation of the phases, the organic phase is washed with a solution of 25 g of sodium bromide in 250 mL of water and dried over 50 g of sodium sulfate (Na_2SO_4). The solution is filtered using 50 mL of CH_2Cl_2 as a rinse, and 275 mL of ethyl ether is added over 15 min producing a heavy crystalline mass. After stirring for 1 hr, the solids are collected by filtration, washed with 250 mL of 1:4 methylene chloride:ethyl ether and dried at 50°C under vacuum for 1 hr to afford 38.1 g (62.9 mmol, 78%) of O-allyl-N-(9-anthracenylmethyl)cinchonidinium bromide as a crystalline yellow solid (Note 8).

C. *Enantioselective Reaction of N-(Diphenylmethylene)glycine tert-Butyl Ester with Methyl Acrylate – (4S)-2-(Benzhydrylidenamino)pentanedioic Acid, 1-tert-Butyl Ester-5-Methyl Ester.* A flame-dried, 200-mL, three-necked flask, equipped with a large magnetic stirrer, 50 mL addition funnel, septum and nitrogen inlet, is charged with 5.0 g (16.9 mmol) of N-(diphenylmethylene)glycine tert-butyl ester (Note 9), 0.95 g (1.7 mmol) of O-allyl-N-(9-anthracenylmethyl)cinchonidinium bromide and 40 mL of anhydrous CH_2Cl_2. The mixture is cooled in a dry ice-acetone bath to –78°C and 29 g (170 mmol) of cesium hydroxide monohydrate ($CsOH{\bullet}H_2O$) is added to the well-stirred mixture (Note 10). After stirring for 5 min, a solution of 5 mL of methyl acrylate (Note 11) in 10 mL of CH_2Cl_2 is added dropwise over 10 min through the addition funnel. The mixture is stirred vigorously for 3 hr, at which point TLC analysis (Note 12) indicates complete reaction. The mixture is diluted with 150 mL of ethyl ether and 50 mL of water and the cooling bath is removed. The mixture is stirred for 5 min, transferred to a separatory funnel, and 650 mL of ethyl ether is added. The organic phase is washed twice with 200 mL of water and once with 50 mL of brine, dried over magnesium sulfate, filtered, and concentrated on a rotary evaporator. The residue is loaded onto a 6-cm column wet-packed with 80 g of silica gel and hexane. Elution with 1 L of 20:1 hexane:ethyl acetate followed by 1 L of 5:1 hexane:ethyl acetate affords 5.8 g of product as a colorless syrup (90% yield, 95.2% ee for the S-enantiomer, Note 13). Recovery of the catalyst as the chloride salt is accomplished by extraction of the aqueous washes with two 50-mL portions of CH_2Cl_2, drying over Na_2SO_4, filtration, and concentration. Trituration of the residue with ethyl ether affords 0.89 g (94% recovery) of the catalyst as a light yellow solid.

2. Notes

1. The cinchona alkaloids and their quaternary salts are photosensitive and should be stored in brown bottles.

2. (-)-Cinchonidine (96%) was purchased from the Aldrich Chemical Co., Inc. and used as received. It typically contains ~5% of the dihydro analog which behaves like cinchonidine throughout this procedure.

3. 9-Chloromethylanthracene (98+%) was purchased from the Aldrich Chemical Co., Inc. and used as received.

4. The toluene was stored over 4 Å molecular sieves prior to use.

5. The product is isolated as a toluene solvate (mp 154-156°C) and is sufficiently pure for use in the following step. A reference sample of the solvent-free salt can be prepared by dissolving the toluene solvate in CH_2Cl_2 (1 g/5 mL) followed by filtration, washing and drying of the resulting crystals of N-(9-anthracenylmethyl)cinchonidinium chloride: $[\alpha]_D^{23}$ –363 (*c* 0.81, $CHCl_3$); mp 168-171°C (dec.); FTIR (film) cm^{-1}: 3500-2500, 1625, 1589, 1571, 1509, 1478, 1462, 1451, 1422, 1265, 1062; ^1H NMR (400 MHz, $CDCl_3$): δ 0.99 (m, 1H), 1.09 (m, 1H), 1.68 (bs, 1H), 1.80 (m, 2H), 2.12 (bs, 1H), 2.40 (app. t, 1H, J = 11.1), 2.56 (dd, 1H, J = 10.7, 12.8), 4.07 (bd, 1H, J = 12.9), 4.71 (m, 2H), 4.89 (dd, 1H, J = 1.3, 10.5), 5.25 (dd, 1H, J = 1.0, 17.3), 5.42 (m, 1H), 6.67 (d, 1H, J = 13.6), 6.83 (d, 1H, J = 13.5), 7.24 (m, 4H), 7.20 (m, 2H), 7.38 (m, 1H), 7.56 (d, 1H, J = 8.2), 7.60 (m, 1 H), 7.63 (d, 1H, J = 8.2), 7.96 (s, 1H), 8.02 (d, 1H, J = 4.4), 8.21 (d, 1H, J = 5.2), 8.71 (d, 1H, J = 8.2), 8.85 (m, 2H), 9.06 (d, 1H, J = 9.0); ^{13}C NMR (100 MHz, $CDCl_3$): δ 23.4, 25.7, 25.9, 38.5, 50.4, 54.8, 61.3, 67.0, 67.3, 117.7, 118.3, 120.1, 124.1, 124.2, 124.7, 124.8, 125.6, 126.3, 126.9, 127.4, 127.6, 128.3, 128.5, 128.6, 129.2, 130.2, 130.4, 131.1, 132.7, 133.2, 136.4, 145.7, 147.1, 149.4; FABMS: 485 [M-Cl]$^+$; HRMS calcd for $[C_{34}H_{33}N_2OCl-Cl]^+$: 485.2593, found: 485.2575. Anal. Calcd for

$C_{34}H_{33}ClN_2O$: C, 78.37; H, 6.38; Cl, 6.80; N, 5.38. Found: C, 78.04; H, 6.42; Cl, 6.81; N, 5.50.

6. Allyl bromide (99%) was purchased from the Aldrich Chemical Co., Inc. and used as received. Excess reagent is used, as hydrolysis of the allyl bromide is competitive with the O-alkylation of N-(9-anthracenylmethyl)cinchonidinium chloride.

7. The 50% (w/w) aqueous potassium hydroxide (KOH) solution is prepared immediately before use in a neoprene bottle. *Caution: Cooling with a water bath is necessary since dissolution of KOH is very exothermic.*

8. O-Allyl-N-(9-anthracenylmethyl)cinchonidinium bromide has the following properties: $[\alpha]_D^{23}$ -320 (*c* 0.45, $CHCl_3$); mp 194-197°C; FTIR (film) cm^{-1}: 3504, 3082, 2950, 2907, 2884, 1646, 1641, 1625, 1588, 1509, 1450, 1067, 996; ^1H NMR (400 MHz, CD_3OD); δ 1.60 (m, 2H), 1.96 (d, 1H, J = 2.9), 2.17 (m, 1H), 2.48 (m, 2H), 2.61 (m, 2H), 7.79-7.77 (m, 2H), 7.95-7.92 (m, 3H), 8.25-81.9 (m, 3H), 8.44 (d, 1H, J = 9.0), 8.57 (m, 1H), 8.76 (d, 1H, J = 9.0), 8.89 (s, 1H), 9.02 (d, 1H, J = 4.6); ^{13}C NMR (100 MHz, CD_3OD): δ 23.4, 26.2, 27.3, 39.5, 49.9, 53.6, 57.4, 63.4, 69.9, 71.4, 117.8, 119.0, 121.8, 125.7 (2C), 126.5, 126.6, 127.1, 129.2, 129.5, 130.5, 131.1, 131.3, 131.5, 133.0, 133.1, 133.8, 133.9, 134.6, 134.7, 134.8, 138.6, 143.0, 149.3, 151.1; FABMS: 525 [M-Br]$^-$; HRMS calcd for $[C_{37}H_{37}N_2OBr-Br]^-$: 525.2906, found: 525.2930. Anal. Calcd for $C_{37}H_{37}BrN_2O$: C, 73.38; H, 6.16; Br, 13.19; N, 4.63. Found: C, 73.40; H, 6.12; Br, 13.19; N 4.47.

9. N-(Diphenylmethylene)glycine tert-butyl ester was purchased from the Aldrich Chemical Co., Inc. and used as received.

10. Cesium hydroxide monohydrate was purchased from the Aldrich Chemical Co., Inc. and used as received.

11. Methyl acrylate was purchased from the Aldrich Chemical Co., Inc. and used as received.

12. The TLC analysis was performed using silica gel plates, 10% ethyl acetate in hexane and UV detection.

13. Characterization data for the Michael adduct: α_D^{23} -100 (*c* 1.35, CH_2Cl_2); mp 194-196°C; FTIR (film) cm^{-1}: 3061, 3056, 2977, 2950, 2933, 1735, 1624, 1446, 1368, 1316, 1277, 1254, 1195, 1150; ^1H NMR (400 MHz, $CDCl_3$): δ 1.43 (s, 9H), 2.24-2.18 (m, 2H), 2.39-2.35 (m, 2H), 3.59 (s, 3H), 3.95 (dd, 1H, J = 7.2, 5.6), 7.18-7.16 (m, 2H), 7.45-7.25 (m, 6H), 7.63 (d, 2H, J = 7.2); ^{13}C NMR (125 MHz, $CDCl_3$): δ 28.3, 28.8, 30.7, 51.8, 65.0, 81.5, 128.1, 128.3, 128.7, 128.9, 129.1, 130.6, 136.6, 139.6, 170.5, 170.9, 173.8; CIMS: 382 $[M+H]^+$, 280, 134; HRMS calcd for $[C_{23}H_{27}NO_4+H]^+$: 382.2018, found: 382.2017. The enantioselectivity was determined by chiral HPLC analysis (Regis Whelk-O1 column, 20% 2-propanol-hexane, 0.5 mL/min, λ = 254 nm, retention times: R (minor): 16.1 min, S (major): 19.1 min).

Waste Disposal Information

All toxic materials were disposed of in accordance with "Prudent Practices in the Laboratory"; National Academic Press; Washington, DC 1998.

3. Discussion

O-Allyl-N-(9-anthracenylmethyl)cinchonidinium bromide can be used as a versatile phase transfer catalyst for highly enantioselective alkylation and Michael addition reactions of the benzophenone Schiff base of glycine tert-butyl ester.[3,4,5] The reaction is performed with 10 mol% of the cinchona alkaloid catalyst, 1.5 to 5 equivalents of the electrophile and 10 equivalents of $CsOH \cdot H_2O$ in methylene chloride at −60 to −78°C.[3] The reaction may be performed using 50% aqueous KOH (w/v) at 0°C instead of $CsOH \cdot H_2O$ with a slight reduction in enantioselectivity. The catalyst itself is prepared from inexpensive, readily available starting materials in two steps without chromatography. It is stable at room temperature for many months and can be recovered from the asymmetric alkylation reaction and reused. The alkylation reaction can be performed with a variety of electrophiles, including aliphatic, allylic and benzylic halides as well as cyclic and acyclic α,β-unsaturated esters and ketones, affording convenient access to orthogonally protected, enantiomerically pure amino acid derivatives useful as chiral building blocks in asymmetric synthesis. The use of the 9-anthracenylmethyl group on the cationic nitrogen rigidifies the catalyst and favors reaction through a highly structured contact ion pair.[3-5] Similar reactions using previously prepared catalysts are less enantioselective.[6,7,8]

1. Harvard University, Department of Chemistry and Chemical Biology, 12 Oxford Street, Cambridge, MA 02138.
2. Pfizer Inc., Central Research, Eastern Point Road, Groton, CT 06340.

3. Corey, E. J.; Xu, F.; Noe, M. C. *J. Am. Chem. Soc.* **1997**, *119*, 12414.
4. Corey, E. J.; Bo, Y.; Busch-Petersen, J. *J. Am. Chem. Soc.* **1998**, *120*, 13000.
5. Corey, E. J.; Noe, M. C.; Xu, F. *Tetrahedron Lett.* **1998**, *39*, 5347.
6. O'Donnell, M. J.; Bennett, W. D.; Wu, S. *J. Am. Chem. Soc.* **1989**, *111*, 2353.
7. O'Donnell, M. J.; Wu, S.; Huffman, J. C. *Tetrahedron* **1994**, *50*, 4507.
8. Lipkowitz, K. B.; Cavanaugh, M. W.; Baker, B.; O'Donnell, M. J. *J. Org. Chem.* **1991**, *56*, 5181.

Appendix

Chemical Abstracts Nomenclature (Collective Index Number)

(Registry Number)

O-Allyl-N-(9-anthracenylmethyl)cinchonidinium bromide: Cinchonanium, 1-(9-anthracenylmethyl)-9-(2-propenyloxy)-, bromide, (8α,9R)- (9); (200132-54-3)

Cinchonidine: Cinchonan-9-ol, (8α, 9R)- (9); (485-71-2)

9-Chloromethylanthracene: Anthracene, 9-(chloromethyl)- (8,9); (24463-19-2)

Allyl bromide:1-Propene, 3-bromo- (9); (106-95-6)

N-(Diphenylmethylene)glycine tert-butyl ester: Glycine, (diphenylmethylene)-, 1,1-dimethylethyl ester (9); (81477-94-3)

Methyl acrylate: 2-Propenoic acid, methyl ester (9); (96-33-3)

(4S)-2-(Benzhydrylidenamino)pentanedioic acid, 1-tert-butyl ester-5-methyl ester: L-Glutamic acid, N-(diphenylmethylene)-, 1-(1,1-dimethylethyl) 5-methyl ester (9); (212121-62-5)

SYNTHESIS OF (R,R)-4,6-DIBENZOFURANDIYL-2,2'-BIS (4-PHENYLOXAZOLINE) (DBFOX/PH) – A NOVEL TRIDENTATE LIGAND

[Oxazole, 2,2'-(4,6-dibenzofurandiyl)bis(4,5-dihydro-4-phenyl-, (4R,4'R)-]

A. 1 → 2: 1. s-BuLi, TMEDA, Et_2O, 25°C, 24 hr; 2. CO_2 (g), -78°C; 3. H_2O

B. 2 → 3: $SOCl_2$, $CHCl_3$, Δ, 3 hr

C. 3 → 4: (R)-phenylglycinol, Et_3N, $CHCl_3$, 25°C, 24 hr

D. 4 → 5: 2.3 eq DAST, CH_2Cl_2, -20°C, 18 hr

Submitted by Ulrich Iserloh, Yoji Oderaotoshi, Shuji Kanemasa, and Dennis P. Curran.[1]
Checked by Richard S. Gordon and Andrew B. Holmes.

1. Procedure

A. *Dibenzofuran-4,6-dicarboxylic acid* (**2**).[2] A 1-L, three-necked, round-bottomed flask (Note 1) is fitted with a large magnetic stirbar. Under argon, the flask is charged with 350 mL of dry, freshly distilled diethyl ether, 10.0 g (59.4 mmol) of dibenzofuran (Notes 2, 3) and 27 mL (179 mmol) of tetramethylethylenediamine (TMEDA). The mixture is cooled to –78°C (bath temperature) with a dry ice/acetone bath, and stirred (Note 4) while 137 mL (178 mmol) of sec-butyllithium is added over 1 hr via cannula and syringe (Note 5). After warming the reaction mixture to 25°C, the suspension is stirred for 24 hr. The mixture is cooled again to –78°C under vigorous stirring and gaseous carbon dioxide (CO_2) (Note 6) is introduced over 1 hr; the reaction mixture is then warmed to 25°C over 4 hr under a constant CO_2 stream (Note 7). After decanting the supernatant liquid, a brown solid is isolated by filtration using a Buchner funnel (Note 8). The residue is washed with 300 mL of diethyl ether and then suspended in 200 mL water, acidified with aqueous 2N HCl to pH 3, and stirred for 1 hr. After filtration, the beige solids are washed first with 400 mL of water and then with 200 mL of diethyl ether. The solids are dried under vacuum over phosphorus pentoxide (P_2O_5) for three days to yield 14.5 g (96%) of the title compound **2** (Note 9).

B. *Dibenzofuran-4,6-dicarbonyl chloride* (**3**): A 250-mL, three-necked flask (Note 1), equipped with a magnetic stirbar, condenser, thermometer, and addition funnel, is charged with 5.0 g (19.6 mmol) of diacid **2** and 68 mL of chloroform ($CHCl_3$) (Note 10). At 25°C, 44 mL (600 mmol) of thionyl chloride (Note 2) and 1 drop of dimethylformamide (DMF) are added, followed by heating the mixture at 68°C for 3 hr. After the initial suspension turns into a yellow solution (Note 11), the heating source is removed and the diacid chloride **3** precipitates as a white solid. After cooling the reaction mixture to 25°C, the solid is collected via filtration using a Buchner funnel, washed with $CHCl_3$ (Note 12), and dried in a vacuum desiccator for 15 hr to give 5.12 g (90%) of **3** as a white powder (Note 13).

C. *(R,R)-Dibenzofuran-4,6-dicarboxylic acid bis(2-hydroxy-1-phenylethyl) amide* (**4**): A 250-mL, three-necked flask (Note 1), equipped with a magnetic stirbar, is charged with 5.00 g (17.07 mmol) of diacid chloride **3** and 135 mL of freshly distilled $CHCl_3$. After cooling the suspension to 0°C under nitrogen, a solution of 5.18 g (37.74 mmol) of (R)-phenylglycinol (Note 2) and 5.26 mL (37.74 mmol) of triethylamine (Et_3N) in 25 mL of $CHCl_3$ (Note 14) is added over 2 hr via an addition funnel (Note 15). After the reaction mixture is stirred for 15 hr at 25°C, $CHCl_3$ (100 mL) and solid ammonium chloride (5 g) is added and the suspension is stirred for a further 0.5 hr (Note 16). The suspension is filtered and the residue is washed with $CHCl_3$ (3 x 50 mL). The residue is put aside for suspension in tetrahydrofuran (THF). The combined filtrate and washings are evaporated to dryness under reduced pressure. The resulting residue is purified by two recrystallizations from ethyl acetate/hexane (ca. 9:1) to give the amide **4** (combined yield 5.2 g, 61.0%). The mother liquor from the recrystallizations is chromatographed using silica gel (eluting with ethyl acetate/hexanes initially in a 1:1 ratio and increasing to 4:1, then 1:0) to give more amide **4** (0.6 g, 7.0%). The original solid residue, obtained from the filtration, is suspended in THF (200 mL); the slurry is stirred using a magnetic stirbar for 30 min. The product is filtered. The THF extraction of this residue is repeated (Note 17) and the filtrates are combined. Removal of the solvent under reduced pressure afforded additional amide **4**, which was recrystallized as before (1.7 g, 20.0%). The overall yield of amide **4** is 5.9-7.5 g (11.9-14.9 mmol, 70-88%) (Note 18).

D. *Caution! Diethylaminosulfur trifluoride is a powerful corrosive and reacts violently with water. Appropriate precautions should be taken whenever it is handled, since one of its decomposition products is hydrofluoric acid.*

(R,R)-4,6-Dibenzofurandiyl-2,2'-bis(4-phenyloxazoline), DBFOX/Ph (**5**):

A 100-mL, one-necked flask (Note 1) equipped with a magnetic stirbar is charged with 2.40 g (4.85 mmol) of **4** and 50 mL of dichloromethane (CH_2Cl_2). After the suspension is cooled (Note 19) to –20°C, 1.49 mL (11.28 mmol) of diethylaminosulfur trifluoride (DAST, Note 2) is slowly added via syringe. After 18 hr at –20°C, 4 mL of aqueous 4N ammonium hydroxide (NH_4OH) is added and the resulting solution is stirred for 15 min at –20°C. Subsequently, the cooling bath is removed and 10 mL of water is added. The orange aqueous phase is extracted with CH_2Cl_2 (3 x 15 mL) and the combined organic phases are dried over magnesium sulfate ($MgSO_4$) and concentrated under vacuum. The resulting material is chromatographed using silica gel (20% EtOAc/hexanes, Note 20) to yield 2.10 g (4.59 mmol, 94%) of **5** as an off-white foam. This foam can be recrystallized from EtOAc/hexanes to yield white crystals (Notes 21- 23).

2. Notes

1. The checkers dried all glassware at 120°C for at least 2 hr, and cooled under argon where appropriate. The submitters dried all glassware at 140°C for at least 4 hr and cooled to 25°C under a flow of argon. Solvents were distilled from the indicated drying agent and were used immediately: dichloromethane (CaH_2), diethyl ether and tetrahydrofuran (sodium/benzophenone), chloroform (potassium carbonate), and triethylamine (KOH).

2. Dibenzofuran (99%), sec-butyllithium (1.3 M in cyclohexane), N,N,N',N'-tetramethylethylenediamine (99%), N,N-dimethylformamide (anhydrous, 99.8%), (R)-(–)-2-phenylglycinol (98%) and (diethylamino)sulfur trifluoride (DAST) were purchased from Aldrich Chemical Company, Inc. Thionyl chloride (99.5%) was purchased from Acros Organics and carbon dioxide (commercial grade) was purchased from Matheson Gas Products Inc. Except for dibenzofuran, all reagents were used as received.

3. Commercial dibenzofuran was not purified. The checkers used recrystallized and commercial dibenzofuran as received in separate experiments with no significant difference in yield.

4. The submitters used a large stirbar followed by a mechanical stirrer (upon introduction of CO_2) to ensure efficient stirring throughout. After initial experiments, the checkers carried out the whole procedure using a powerful magnetic stirrer (and a large magnetic stirbar) with no adverse effect on yield. The checkers observed no improvement in yield when a half-scale reaction was carried out using a mechanical stirrer.

5. To transfer 137 mL of sec-butyllithium, the contents of an entire 100-mL Sure/Seal™ bottle were transferred via cannula, while the remaining 37 mL were transferred via syringe from another bottle.

6. Carbon dioxide was dried by passing the gas through a gas-washing bottle (equipped with a fritted cylindrical gas-dispersion tube) containing concentrated sulfuric acid. For safety reasons, empty gas-washing bottles were placed before and after the H_2SO_4-containing bottle. The gas was introduced into the flask through a wide pipette (0.8 cm i. d.) which was submerged below the level of the reaction mixture. The checkers observed very little difference in yield irrespective of whether the inlet tube was above or below the surface of the reaction mixture.

7. The beige-colored suspension turned white upon introduction of carbon dioxide. The submitters adjusted the rate of addition such that the internal temperature remained below –65°C. At –25°C, the white suspension turned red-brown. The checkers did not monitor the internal temperature of the reaction, but ensured that the CO_2 addition was slow and constant.

8. The yellow supernatant/filtrate comprises TMEDA in Et_2O, while the remaining solid is the lithiocarboxylate of diacid **2**. No precaution was taken during filtration to exclude air or moisture.

9. The submitters obtained 15.1 g (99%). Data for **2**: ^1H NMR (250 MHz, DMSO-d_6) δ 7.53 (dd, 2 H, J = 7.5 and 7.5), 8.04 (d, 2 H, J = 7.5), 8.44 (d, 2 H, J = 7.5), 13.30 (bs, 2 H); ^{13}C NMR (100 MHz, DMSO-d_6) δ 116.9, 123.7, 124.9, 126.2, 130.4, 154.6, 165.7; IR (solid phase) cm^{-1}: 2858, 1686, 1431, 1305, 1288, 1154; MS (EI) m/z 256 (M^+, 15), 181 (15), 69 (100); HRMS (ES, M + Na^+) m/z Calcd for $C_{14}H_8O_5Na$ 279.0264. Found 279.0269, Anal. Calcd for $C_{14}H_8O_5$: C, 65.6; H, 3.1. Found: C, 65.1; H, 3.2. The submitters reported mp 324-325°C. The checkers could not obtain a mp for this material, but observed decomposition at temperatures > 320°C (lit.[4] > 300°C).

10. The checkers carried out the reaction with undistilled chloroform, with no difference in yields.

11. Although the solution is yellow, the reaction mixture remained cloudy throughout.

12. It is important not to use too much chloroform because the product is sparingly soluble.

13. The submitters obtained 5.0 g (87%). Data for **3**: mp 242-244°C, ^1H NMR (400 MHz, $CDCl_3$) δ 7.59 (dd, 2 H, J = 5 and 5), 8. 31 (dd, 2 H, J = 5 and 1), 8.33 (dd, 2 H, J = 5 and 1); ^{13}C NMR (100 MHz, DMSO-d_6) δ 119.3, 123.9, 124.8, 127.7, 132.6, 163.4, 168.2; IR (solid phase) cm^{-1}: 1757, 1622, 1587, 1474, 1425, 1236, 1184, 1142; MS (EI) m/z 292 (M^+, 40), 257 (40), 169 (20), 119 (20); HRMS (EI, M^+) m/z Calcd for $C_{14}H_6Cl_2O_3$ 291.9694. Found: 291.9699; Anal. Calcd for $C_{14}H_6Cl_2O_3$: C, 57.5; H, 2.1. Found: C, 56.7; H, 2.1.

14. The checkers had to carefully warm the contents in the addition funnel (using a heat gun) to completely dissolve the (R)-(–)-2-phenylglycinol. The contents of the funnel were allowed to cool before addition to the reaction mixture.

15. The submitters observed that slow addition is necessary to avoid formation of a thick gel-like slurry and noted that adding additional $CHCl_3$ is beneficial if the slurry[3] is too viscous to stir. The submitters observed (over the course of 15 hr) the formation of a thick slurry which in turn became a free flowing white suspension. The checkers observed the

immediate formation of a clear yellow solution, which became cloudy, and sometimes a thick slurry, during the course of the reaction.

16. The product is partially soluble in $CHCl_3$ and additional $CHCl_3$ is added to recover as much product as possible. However, significant product still remains undissolved.

17. The submitters omitted the extraction of the first residue (which consists of triethylamine hydrochloride, excess ammonium chloride (NH_4Cl) and amide **4**) with THF, but the checkers found this essential in order to obtain the higher yield (88%) reported. To recover all amide **4**, it is necessary to repeat the procedure, because the product requires extended periods to dissolve completely. As an alternative work up procedure, the checkers removed the $CHCl_3$ under reduced pressure and stirred the reaction mixture in THF (250 mL containing ca. 5 g of NH_4Cl) for 1 hr. After filtration of the suspension, recovery of **4** was poor (3.6 g from two recrystallizations).

18. The submitters obtained 7.4 g (87%) of **4**. Data for **4**: mp 185-186°C; ^{1}H NMR (250 MHz, DMSO-d_6) δ 3.66-3.81 (m, 4 H, C<u>H</u>$_2$OH), 4.98 (t, 2 H, J = 5, CH$_2$O<u>H</u>), 5.20 (dd, 2 H, J = 7 and 14, C<u>H</u>Ph), 7.16-7.27 (m, 6 H), 7.42-7.60 (m, 6 H), 8.00 (dd, 2 H, J = 1 and 8), 8.36 (dd, 2 H, J = 1 and 8), 8.81 (d, 2 H, J = 8, PhCHN<u>H</u>); ^{13}C NMR (63 MHz, DMSO-d_6) δ 56.5, 65.3, 120.1, 123.9, 124.4, 124.7, 127.4, 127.6, 128.2, 128.7, 141.3, 153.1, 163.8; IR (solid phase) cm^{-1}: 3365, 3278, 1636, 1571, 1530, 1339, 1290, 1190, 1029; MS (ES) m/z 517 (M+Na$^+$, 40), 495 (M+H, 100), 375 (50), 255 (25); HRMS (ES, M + Na$^+$) m/z Calcd for $C_{30}H_{26}N_2O_5Na$: 517.1739. Found: 517.1732; Anal. Calcd for $C_{30}H_{26}N_2O_5$: C, 72.9; H, 5.3; N, 6.0. Found: C, 72.5; H, 5.4; N, 5.7.

19. The checkers used an electrical refrigeration bath to maintain this reaction temperature.

20. The silica gel is stirred for 30 min in 20% EtOAc/hexanes containing 5% (v/v) Et_3N prior to packing the column. Before applying the crude material, the column was equilibrated with 20% EtOAc/hexanes to remove excess Et_3N.

21. Data for recrystallized **5**: R_f (**5**) = 0.19 (40% EtOAc/hexanes), mp 126-127°C (submitters observed mp 125.0-126°C, lit.[4] 134.0-135°C), $[\alpha]_D^{23}$ +46.92 (c 0.97, $CHCl_3$) [submitters observed $[\alpha]_D^{23}$ +50.75 (c 1.07, $CHCl_3$)]; [lit.[4] + 47.75 (c 1.07 $CHCl_3$)], ^1H NMR (250 MHz, $CDCl_3$) δ 4.38 (br t, 2 H, J = 8), 4.96 (dd, 2 H, J = 10 and 10), 5.54 (dd, 2 H, J = 9.5 and 9.5), 7.23-7.45 (m, 12 H), 8.10-8.19 (m, 4 H); ^{13}C NMR (63 MHz, $CDCl_3$) δ 69.9, 74.9, 113.1, 123.2, 124.0, 124.9, 126.9, 127.6, 128.8, 142.4, 154.4, 162.4; IR (solid phase) cm^{-1}: 1642, 1424, 1412, 1283, 1184, 1120, 1072; MS (ES) m/z 459 (M+H, 95), 253 (10); HRMS (ES, M + H) m/z Calcd for $C_{30}H_{22}N_2O_3Na$: 459.1708. Found: 459.1721. Anal. Calcd for $C_{30}H_{22}N_2O_3$: C, 78.6; H, 4.8; N, 6.1. Found: C, 78.5; H, 4.8; N, 6.1.

22. The checkers also prepared (S,S)-**5** over two steps according to the same procedure from the reaction of **3** with (S)-(–)-2-phenylglycinol purchased from Aldrich Chemical Co., Inc. In addition to the spectroscopic properties of (S,S)-**5** which were identical to those reported above for the (R,R)-enantiomer, the checkers observed mp 126.0-127°C, $[\alpha]_D^{23}$ -43.20 (c 1.00, $CHCl_3$).

23. The enantiomeric purities of (R,R)-**5** and (S,S)-**5** were measured separately by courtesy of Mr. Eric Hortense (GlaxoSmithKline, Stevenage). Chiral HPLC analysis [25 cm Chiracel OD-H, column no. ODHOCE-IF029, mobile phase ethanol/heptane 1:4 (v/v), UV detector at 215 nm, flow rate 1.0 mL/min at 25°C] of (R,R)-**5** yielded a retention time of 7.4 min, while that of enantiomer (S,S)-**5** showed a retention time of 8.7 min. The enantiomeric ratio of (R,R)-**5** was in excess of 99.5:0.5, while that of (S,S)-**5** was also >99.5:0.5. The submitters determined the enantiomeric purity of (R,R)-5 as 99% ee by chiral HPLC on a Chiracel OD column (7% 2-propanol/hexanes, 1 mL/min, t(R) = 45 min, t(S) = 65 min).[4]

Waste Disposal Information

All toxic materials were disposed of in accordance with "Prudent Practices in the Laboratory"; National Academic Press; Washington, DC, 1995.

3. Discussion

(R,R)-4,6-Dibenzofurandiyl-2,2'-bis(4-phenyloxazoline), DBFOX/Ph (5), is a novel tridentate bisoxazoline ligand developed by Kanemasa and coworkers[5] that has been successfully used as a chiral Lewis acid in enantioselective Diels-Alder-reactions, nitrone cycloadditions and conjugate additions of radicals[6] and thiols[7] to 3-(2-alkenoyl)-2-oxazolidinones. Representative examples for cycloadditions using the $Ni(ClO_4)_2 \cdot 6H_2O$ derived complex are shown below.

The procedure described here is a variant of the original preparation[5] that offers the following advantages:

(1) The overall yield of the four-step ligand preparation is increased from 28% to 71%. The improved dicarboxylation of 4,6-dilithiodibenzofuran significantly contributes to this yield increase. Originally, the dilithiated dibenzofuran was quenched with dry ice (solid CO_2), which presumably led to competitive protonation and ketone formation. In this procedure, dry CO_2 is introduced as a gas and gives a quantitative yield of dicarboxylated material **2**.

(2) The original protocol was not amenable for scale-up. The current procedure performs reliably on a 2-5 gram scale.

(3) All intermediates are purified by recrystallization. This provides for clean bis(2-hydroxy-1-phenylethyl) amide **4**, which is essential for the excellent yields obtained in the final cyclodehydration.

(4) The DAST-mediated cyclodehydration gives higher yields of DBFOX than the original base-mediated cyclization.

1. Department of Chemistry, University of Pittsburgh, Pittsburgh, PA 15260.

2. We acknowledge Professor X. Zhang, The Pennsylvania State University, for helpful suggestions and we also thank Dr. Ana Gabarda for helping with Step A.
The dilithiation of dibenzofuran has been previously reported: (a) Haenel, M. W.; Jakubik, D.; Rothenberger, E.; Schroth, G. *Chem. Ber.* **1991**, *124*, 1705; (b) Jean, F.; Melnyk, O.; Tartar, A. *Tetrahedron Lett.* **1995**, *36*, 765.

3. Gel formation was also observed by other researchers: Takacs, J. M.; Quincy, D. A.; Shay, W.; Jones, B. E.; Ross, C. R. *Tetrahedron: Asymmetry* **1997**, *8*, 3079.

4. Kanemasa, S.; Oderaotoshi, Y.; Sakaguchi, S.; Yamamoto, H.; Tanaka, J.; Wada, E.; Curran, D. P. *J. Am. Chem. Soc.* **1998**, *120*, 3074.

5. Kanemasa, S.; Oderaotoshi, Y.; Tanaka, J.; Wada, E. *J. Am. Chem. Soc.* **1998**, *120*, 12355.

6. Iserloh, U.; Curran, D. P.; Kanemasa, S. *Tetrahedron: Asymmetry* **1999**, *10*, 2417.

7. Kanemasa, S.; Oderaotoshi, Y.; Wada, E. *J. Am. Chem. Soc.* **1999**, *121*, 8675.

Appendix

Chemical Abstracts Nomenclature (Collective Index Number);

(Registry Number)

(R,R)-4,6-Dibenzofurandiyl-2,2'-bis(4-phenyloxazoline) (DBFOX/PH): Oxazole, 2,2'-(4,6-dibenzofurandiyl)bis(4,5-dihydro-4-phenyl-, (4R,4'R- (9); (195433-00-2)

Dibenzofuran: (8,9); (132-64-9)

sec-Butyllithium: Lithium, (1-methylpropyl)- (9); (598-30-1)

N,N,N',N'-Tetramethylethylenediamine, 1,2-Ethanediamine, N,N,N',N'-tetramethyl- (9); (110-18-9)

Carbon dioxide: (8,9); (124-38-9)

Dibenzofuran-4,6-dicarboxylic acid: 4,6-Dibenzofurandicarboxylic acid (9); (88818-47-7)

Thionyl chloride (8,9); (7719-09-7)

N,N-Dimethylformamide: Formamide, N,N-diethyl- (8,9); (68-12-2)

Dibenzofuran-4,6-dicarbonyl chloride: 4,6-Dibenzofurandicarbonyl dichloride (9); (151412-73-8)

(R)-(--)-2-Phenylglycinol: Benzeneethanol, β-amino-, (βR)- (9); (56613-80-0)

Triethylamine: Ethanamine, N,N-diethyl- (9); (121-44-8)

Diethylaminosulfur trifluoride (DAST): Sulfur, (N-ethylethanaminato)trifluoro-, (T-4) (9); (38078-09-0)

(R,R)-Dibenzofuran-4,6-dicarboxylic acid bis(2-hydroxy-1-phenylethyl) amide: 4,6-Dibenzofurandicarboxamide, N,N'-bis[(1R)-2-hydroxy-1-phenylethyl]- (9); (247097-79-6)

(4S)-4-(1-METHYLETHYL)-5,5-DIPHENYL-2-OXAZOLIDINONE

(2-Oxazolidinone, 4-(1-methylethyl)-5,5-diphenyl-, (4S)-)

A. MeI, $KHCO_3$ / DMF → **1**

B. PhBr + Mg → THF → PhMgBr

C. 2 PhMgBr + **1** → THF, then H_2O → **2**

D. **2** → KO*t*-Bu / THF → **3**

Submitted by Meinrad Brenner,[1] Luigi La Vecchia,[2] Thomas Leutert,[2] and Dieter Seebach.[1]

Checked by Mitsuru Kitamura and Koichi Narasaka.

1. Procedure

A. *N-(tert-Butoxycarbonyl)-L-valine methyl ester* (**1**) (Note 1). A 1-L, three-necked flask, fitted with a pressure-equalizing, 50-mL addition funnel, thermometer, and magnetic stirring bar is charged under an argon atmosphere (Note 2) with (S)-N-(tert-butoxycarbonyl)valine (65.2 g, 0.300 mol) (Note 3), dimethylformamide (DMF)

(440 mL) (Note 4) and potassium hydrogen carbonate (60.1 g, 0.600 mol) (Note 5). Methyl iodide (29.9 mL, 0.480 mol) (Note 6) is added dropwise to this suspension over a period of 30 min. After the addition is complete, stirring is continued for 3 hr. The reaction mixture is poured into a 2-L separatory funnel, diluted with water (1.2 L) and extracted with a 1:1 mixture of ethyl acetate and hexane (3 × 250 mL). The organic phases are combined and successively washed with water (2 × 250 mL), aqueous 5% sodium sulfite (2 × 250 mL), and brine (250 mL). After drying over anhydrous magnesium sulfate, the solvent is removed and the oily residue is dried under vacuum (10 mbar) at room temperature. Crude **1** (69.4 g) is used in Step C without further purification.

B. *Phenylmagnesium bromide.* Under an argon atmosphere (Note 2) a 2-L, three-necked flask, fitted with a reflux condenser, pressure-equalizing, 500-mL addition funnel, thermometer, and magnetic stirring bar, is charged with magnesium turnings (25.5 g, 1.05 mol) and tetrahydrofuran (THF) (130 mL) (Note 7). Bromobenzene (15.7 g, 0.100 mol) (Note 8) is then added, and the reaction is initiated by warming with a heat gun. A solution of bromobenzene (149 g, 0.95 mol) in THF (500 mL) is added at such a rate that gentle refluxing is maintained. After the addition is complete (Note 9), the reaction mixture is heated to reflux for 1 hr with an oil bath.

C. *(S)-[1-(Hydroxydiphenylmethyl)-2-methylpropyl]carbamic acid, tert-butyl ester* (**2**). The Grignard solution obtained in Step B is cooled with an ice bath (Note 10) and a solution of N-(tert-butoxycarbonyl)-L-valine methyl ester (**1**) (69.4 g, 0.30 mol) in THF (300 mL) is added at a rate to maintain the internal temperature below 6°C (Note 11). After the addition is complete, the reaction mixture is allowed to warm to room temperature and stirring is continued for another 15 hr (Note 12).

The reaction mixture is poured into an ice-cold aqueous saturated ammonium chloride solution (700 mL). The aqueous phase is separated and extracted with ethyl acetate (2 × 500 mL). The organic phases are combined, washed with brine (500 mL), and dried over anhydrous magnesium sulfate. After removal of the solvent under reduced pressure, crude **2** (115 g) is obtained as a white solid.

The crude product is dissolved in boiling ethyl acetate (600 mL) (Note 13) in a 1-L flask fitted with a reflux condenser and a magnetic stirring bar; then hexane (230 mL) is added. A few minutes after cessation of heating, a white solid starts to precipitate. The suspension is allowed to cool to room temperature, then is cooled for an additional hour with an ice bath. The solids are collected, washed with ice-cold hexane (100 mL), and dried under high vacuum (0.1 mbar) at room temperature for 12 hr to give purified **2** (77.1 g, 72 %) (Note 14) as a white solid. The mother liquor is concentrated under reduced pressure (to 220 g) and hexane (150 mL) is added. A few minutes after the addition of hexane, colorless crystals start to form. After standing for several hours at room temperature, the crystals are collected, washed with ice-cold hexane (100 mL) and dried as described above to yield a second crop of **2** (12.5 g, 12%). A third crop (11.3 g, 11%) is obtained after concentration (to 70 g) and treating of the mother liquor as described above (Note 15).

D. *(4S)-4-(1-Methylethyl)-5,5-diphenyl-2-oxazolidinone* (**3**). Under an argon atmosphere (Note 2), a 2-L, three-necked flask, fitted with a thermometer and magnetic stirring bar (Note 16) is charged with (S)-[1-(hydroxydiphenylmethyl)-2-methylpropyl]carbamic acid, tert-butyl ester (**2**) (97.8 g, 0.275 mol) (Note 17) and THF (1.8 L). The resulting solution is cooled to an internal temperature below 5°C with an ice bath. Potassium tert-butoxide (37.0 g, 0.330 mol) (Note 18) is then added in one portion (Note 19). A few minutes after the addition of potassium tert-butoxide a white solid begins to precipitate. Stirring below 5°C is continued for 2 hr. The resulting suspension is poured into a 10% aqueous solution of ammonium chloride (2 L) and stirred for 10 min. The white solids are collected and washed with water (4 × 400 mL) (Note 20). The solids are transferred to a 1-L flask fitted with a reflux condenser and a magnetic stirring bar. Methanol (650 mL) is added and the resulting suspension is heated to reflux for 1 hr with stirring. The mixture is allowed to cool to room temperature and stirring is continued for 1 hr in an ice bath. The white solids are collected and washed with ice-cold methanol (100 mL). After drying for 24 hr under high vacuum (0.1 mbar) at room temperature, oxazolidinone **3** (72.5 g, 94%) (Note 21) is obtained as a white solid.

2. Notes

1. N-(tert-Butoxycarbonyl)-L-valine methyl ester is also commercially available (Aldrich Chemical Co., Inc). The procedure described here is taken from the literature[3] and is readily reproducible.

2. The flask was flushed with argon and the argon atmosphere was maintained with an argon balloon.

3. (S)-N-(tert-Butoxycarbonyl)valine (puriss grade) was obtained from Fluka Chemie AG. (the submitters) or from Tokyo Chemical Industry (the checkers).

4. DMF (p. a. grade) was used as received from Merck AG. The checkers obtained DMF (99%) from Kokusan Chemical Co., Ltd.

5. Powdered potassium hydrogen carbonate (purum grade) was purchased from Fluka Chemie AG (the submitters) or Kokusan Chemical Co., Ltd. (the checkers).

6. Methyl iodide (purum grade) was obtained from Fluka Chemie AG (the submitters) or Tokyo Chemical Industry (the checkers).

7. THF (p. a. grade) was used as received from Merck AG. The checkers obtained THF from Kanto Chemical Co. (<0.005% water).

8. Bromobenzene (puriss grade) was obtained from Fluka Chemie AG (the submitters) or Tokyo Chemical Industry (the checkers).

9. The bromobenzene solution is added over ca. 1.5 hr.

10. The clear dark Grignard solution becomes a suspension after cooling with an ice bath.

11. The protected valine ester is added over ca. 1.5 hr.

12. TLC analysis indicates almost complete reaction after stirring for 3 hr at room temperature. Stirring is continued overnight to ensure complete reaction.

13. The resulting light yellow solution is slightly turbid.

14. The product exhibits the following properties: mp 188.5-189°C. $[\alpha]_D^{r.t.}$ = -65 (c = 1.0, CH_2Cl_2) [The checkers obtained $[\alpha]_D^{29°C}$ = -60 (1.0, CH_2Cl_2)]. IR ($CHCl_3$) cm^{-1}: 3446, 2965, 1703, 1498, 1368. ^{1}H NMR (400 MHz, $CDCl_3$) δ: 0.88 (d, J = 6.8, Me), 0.90 (d, J = 6.9, Me), 1.32 (s, t-Bu), 1.43 (br. s, t-Bu), 1.80 (septd, J = 6.8, 2.2, CH), 2.63 (s, OH), 4.42-4.50 (m, CH), 4.58-4.62 (m, CH), 5.01 (d, J = 10.2, NH), 7.13-7.32 (m, 6 H), 7.44-7.55 (m, 4 H). ^{13}C NMR (100 MHz, $CDCl_3$) δ: 17.4, 22.7, 28.3, 28.8, 59.1, 79.0, 82.4, 125.3, 125.7, 126.7, 126.8, 128.2, 128.3, 145.6, 146.3, 156.3. Exact mass calc. for $C_{22}H_{29}NO_3Na^+$ (M + Na^+): 378.2040. Found (MALDI) m/z 378.2042. For comparison with literature data, see Ref. 4 and 5.

15. The third crop contains (4S)-4-(1-methylethyl)-5,5-diphenyloxazolidin-2-one (**3**) as a byproduct (ca. 7 mol% as determined from the ^{1}H NMR spectrum).

16. Since the reaction mixture becomes a viscous suspension, a large magnetic stirring bar or mechanical stirrer has to be used.

17. The three crops obtained in Step B are used.

18. Potassium tert-butoxide (purum grade) was purchased from Fluka Chemie AG.

19. The temperature rises to 8°C after the addition of potassium tert-butoxide.

20. The third and fourth washes are negative when tested for chloride with silver nitrate.

21. The enantiomer ratio was determined by HPLC (column: Chiralcel OD-H, 4.6 × 15 mm, 5 μm, Daicel Chemical Industries; eluent: hexane/2-PrOH 10:1; flow: 1 mL/min; detection: UV at 254 nm; (S)-**3**: t_R = 4.3 min; (R)-**3**: t_R = 12.8 min) to be >99:1. The product exhibits the following properties: mp 253.5-254°C, $[\alpha]_D^{r.t}$ = -244 (c = 0.61, CH_2Cl_2). IR ($CHCl_3$) cm^{-1}: 3460, 2967, 1759. ^{1}H NMR (400 MHz, $CDCl_3$) δ:

0.69 (d, J = 6.6, Me), 0.90 (d, J = 7.0, Me), 1.81-1.90 (m, CH), 4.36 (d, J = 3.7, CH), 6.61 (br. s, NH), 7.23-7.40 (m, 8 H), 7.52-7.55 (m, 2 H). ^{13}C NMR (100 MHz, $CDCl_3$) δ: 15.6, 20.8, 29.6, 65.9, 89.4, 125.7, 126.3, 127.7, 128.1, 128.2, 128.5, 139.2, 143.9, 158.8. Exact mass calc. for $C_{18}H_{19}NO_2Na^+$ (M + Na^+): 304.1308. Found (MALDI) m/z 304.1308. For comparison with literature data see Ref. 5–8.

Waste Disposal Information

All toxic materials were disposed of in accordance with "Prudent Practices in the Laboratory"; National Academy Press; Washington, DC, 1995.

3. Discussion

The present procedure for the preparation of oxazolidinone **3** is a variation of the procedures described by Luche[4] and Davies.[5] Yields have been substantially enhanced by improving the purification procedures. The preparation of **3** starting from valine methyl or ethyl ester hydrochloride has been described by several authors.[6–9] These procedures suffer from moderate yields for the Grignard addition step and some of them use hazardous reagents like phosgene.

Oxazolidinones, such as **4** and **5**, have been introduced by Evans as very useful chiral auxiliaries.[10] These compounds are prepared from readily available chiral amino alcohols and are successfully employed in a variety of different stereoselective reactions.[11]

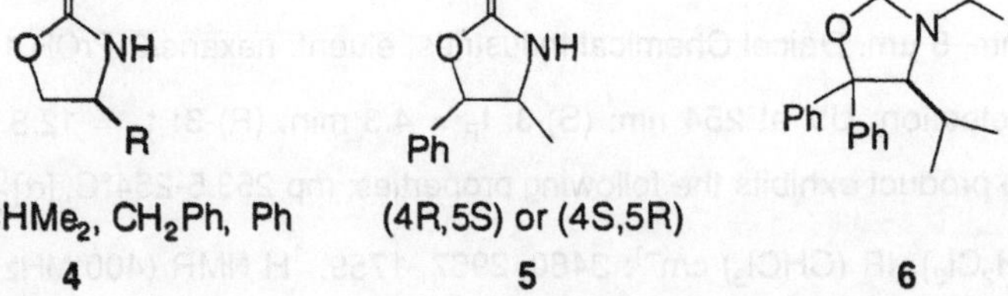

R = $CHMe_2$, CH_2Ph, Ph (**4**)   (4R,5S) or (4S,5R) (**5**)   **6**

(4S)-4-(1-Methylethyl)-5,5-diphenyl-2-oxazolidinone (**3**), whose preparation is described here, has several advantages over Evans' original auxiliaries:[8] i) Derivatives of **3** are more likely to crystallize. In many cases the separation and purification of diastereoisomers can be achieved by simple recrystallization rather than by expensive and time-consuming chromatography. ii) Acylation of **3** can be carried out at 0°C (instead of –78°C for **4** and **5**) by deprotonation with BuLi, followed by treatment with an activated carboxylic acid derivative. iii) Lithium enolates of N-acyl derivatives of **3** can be obtained directly by treatment with BuLi at –78°C, in comparison to **4** and **5** when the more expensive lithium diisopropylamide or lithium hexamethyldisilazane is required. iv) The N-acyl derivatives of **3** can be cleaved by using sodium hydroxide without any detectable nucleophilic attack on the oxazolidinone carbonyl group. Expensive and hazardous (in large scale) $LiOH/H_2O_2$ has to be used for the cleavage of N-acyl derivatives of **4** and **5**. v) Due to its low solubility in most solvents, oxazolidinone **3** is easily recovered in high purity by simple filtration after the cleavage step.

Enolates of N-acyl derivatives of oxazolidinone **3** have been used for asymmetric alkylations,[5,7–9] aldol reactions,[8] Mannich reactions,[8] Michael additions to nitroolefins,[8,12] azide transfer reactions,[9] and samarium-Reformatsky reactions.[13] Furthermore, α,β-unsaturated N-acyl derivatives of **3** have been employed in Diels-Alder reactions and in Michael addition reactions of cuprates.[8] The N-methyl-thiomethyl derivative **6** of oxazolidinone **3** can be lithiated to provide a reagent which is synthetically equivalent to a chiral formyl anion.[14]

1. Laboratorium für Organische Chemie, ETH Hönggerberg, Wolfgang-Pauli-Str. 10, HCI, CH-8093 Zürich, Switzerland
2. Novartis Pharma Ltd., Basel, Switzerland.
3. Hamada, Y.; Shibata, M.; Sugiura, T.; Kato, S.; Shioiri, T. *J. Org. Chem.* **1987**, *52*, 1252.

4. Delair, P.; Einhorn, C.; Einhorn, J.; Luche, J. L. *J. Org. Chem.* **1994**, *59*, 4680.
5. Bull, S. D.; Davies, S. G.; Jones, S.; Sanganee, H. J. *J. Chem. Soc., Perkin Trans. 1*, **1999**, *387*.
6. Gawley, R. E.; Zhang, P. *J. Org. Chem.* **1996**, *61*, 8103.
7. Isobe, T.; Fukuda, K. Jap. Pat. JP 09,143,173 (*Chem. Abstr.* **1997**, *127*, 50635x)
8. Hintermann, T.; Seebach, D. *Helv. Chim. Acta* **1998**, *81*, 2093.
9. Gibson, C. L.; Gillon, K.; Cook, S. *Tetrehedron Lett.* **1998**, *39*, 6733.
10. For a review, see: Evans, D. A. *Aldrichimica Acta*, **1982**, *15*, 23.
11. For reviews, see: (a) Ager, D. J.; Prakash, I.; Schaad, D. R. *Chem. Rev.* **1996**, *96*, 835; (b) Ager, D. J.; Prakash, I.; Schaad, D. R. *Aldrichimica Acta*, **1997**, *30*, 3.
12. Brenner, M.; Seebach, D. *Helv. Chim. Acta* **1999**, *82*, 2365.
13. Fukuzawa, S.; Matsuzawa, H.; Yoshimitsu, S. *J. Org. Chem.* **2000**, *65*, 1702.
14. (a) Gaul, C.; Seebach, D. *Org. Lett.* **2000**, *2*, 1501; (b) Gaul, C.; Schärer, K.; Seebach, D. *J. Org. Chem.* **2001**, *66*, 3059; (c) Gaul, C.; Schweizer, B. W.; Seiler, P.; Seebach, D. *Helv. Chim. Acta* **2002**, *85*, 1546.

Appendix

Chemical Abstracts Nomenclature (Collective Index Number); (Registry Number)

(S)-4-(1-Methylethyl)-5,5-diphenyloxazolidin-2-one: 2-Oxazolidinone, 4-(1-methylethyl)-5,5-diphenyl-, (4S)- (9); (184346-45-0)

N-(tert-Butoxycarbonyl)-L-valine methyl ester: Valine, N-[(1,1-dimethylethoxy)carbonyl]-, methyl ester, (S)- (9); (58561-04-9)

(S)-N-(tert-Butoxycarbonyl)valine: L-Valine, N-[(1,1-dimethylethoxy)carbonyl]- (9); (13734-41-3)

Methyl iodide: Methane, iodo- (8, 9); (74-86-4)

Phenylmagnesium bromide: Magnesium, bromophenyl- (8, 9); (100-58-3)

Bromobenzene: Benzene, bromo- (8, 9); (100-86-1)

Magnesium (8, 9); (7439-95-4)

Potassium tert-butoxide: 2-Propanol, 2-methyl-, potassium salt (9); (865-47-4)

(5S)-(d-MENTHYLOXY)-2(5H)-FURANONE

(2(5H)-Furanone, 5[[(1S,2R,5S)-5-methyl-2-(1-methylethyl)cyclohexyl]oxy]-, (5S)-)

A. hν, O_2, CH_3OH, rose bengal

B. CSA, C_6H_6, Δ

C. CSA, CH_2Cl_2, Δ

Submitted by Oscar M. Moradei and Leo A. Paquette.[1]

Checked by Christian Peschko and Rick L. Danheiser.

1. Procedure

A. *5-Hydroxy-2(5H)-furanone.* A solution consisting of freshly distilled furfural (85.84 g, 0.893 mol) (Note 1) and rose bengal (1.82 g, 1.8 mmol, 0.2% mol/eq) (Note 2) in dry methanol (450 mL) is placed in a 600-mL Pyrex photochemical reactor. The outer vessel, which is fitted at its base with a fritted disc, is also equipped with a thermometer, magnetic stirring bar, and reflux condenser. With oxygen bubbling from the bottom of the reactor (Note 3) and appropriate (water) cooling in the reactor well, the stirred solution is irradiated with a tungsten halogen lamp (Note 4). The reaction temperature is maintained at or below 33°C at all times. After 18-22 hr, when no residual furfural can be detected by thin layer chromatography (TLC) (Note 5), the solution is carefully concentrated at 35-38°C (bath temperature) (Note 6). Following standing overnight, the orange solid that forms is taken up in 70 mL of cold (-78°C) chloroform and the resulting crystals are collected by vacuum filtration, washed once with 70 mL of chloroform, and dried under high vacuum to afford 72.70-75.65 g (81-85%) of 5-hydroxy-2(5H)-furanone as sticky, yellowish crystals, mp 52°C (Note 7).

B. *(5S)-(d-Menthyloxy)-2(5H)-furanone.* A 500-mL, round-bottomed flask equipped with a magnetic stirring bar, 10-mL Dean-Stark trap, and reflux condenser is charged with d-menthol (51.9 g, 0.330 mol), 5-hydroxy-2(5H)-furanone (37.3 g, 0.370 mol), D-(+)-camphorsulfonic acid (3.96 g, 0.170 mol), and 190 mL of dry benzene (Note 8). The stirred suspension is heated to reflux under argon with an oil bath preheated to 100°C (Note 9). After 1-2 hr, a total of 5.1 mL of water is collected and no residual menthol is apparent by TLC analysis (Note 10). The reaction mixture is cooled in an ice bath and treated carefully with 100 mL of saturated sodium bicarbonate solution. After completion of the addition, stirring is maintained for 90 min as the mixture is allowed to warm to room temperature. The product is extracted with dichloromethane (300 mL) and the organic phase is washed with three 70-mL portions of brine, dried over anhydrous magnesium sulfate, filtered and concentrated by rotary evaporation to provide 66.7 g (85%) of a 1:1 diastereomeric mixture of the menthyloxybutenolides (Note 11). This mixture is dissolved in 400 mL of hot petroleum ether (bp 35–60°C) and subsequently cooled to room temperature and eventually to –10°C (Note 12). Two additional recrystallizations from petroleum ether (200 mL on each occasion) with slow cooling affords 10.7–17.33 g (14-22%) of pure (5S)-(d-menthyloxy)-2(5H)-furanone, mp 79-80°C (Note 13).

C. *Acid-catalyzed epimerization.* The combined mother liquors from Step B are concentrated to dryness, dissolved in 50 mL of benzene, and concentrated again under reduced pressure. The resulting solid is dissolved in 150 mL of dichloromethane, treated with D-(+)-camphorsulfonic acid (2.0 g, 8.6 mmol) and the mixture is heated to reflux under argon with magnetic stirring for 19 hr. The resulting solution is cooled to 0°C and saturated sodium bicarbonate solution (40 mL) is carefully added. The mixture is stirred for 2 hr at 20°C. The organic phase is separated and washed with two 50-mL portions of water, dried over anhydrous magnesium sulfate, filtered, and concentrated by rotary evaporation. The residue is crystallized three times from petroleum ether as described in Step B to give an additional 13.04–17.60 g (17-22%) of furanone. The total yield of pure (5S)-(d-menthyloxy)-2(5H)-furanone (based on the menthol used in Step B) is 24.75-34.93 g (32-44%) (Note 14).

2. Notes

1. Furfural was purchased from Alfa Aesar and was distilled at reduced pressure (47 mm) from hydroquinone before use.

2. Rose bengal was obtained by the submitters from Acros (certified ≥ 80%) and b y the checkers from the Aldrich Chemical Company, Inc. (certified 89%) and used as received. Anhydrous methanol was obtained from Malinckrodt and used as received.

3. A gentle flow of oxygen is sufficient. A vigorous stream of the gas results in rapid loss of solvent, followed by an undesirable increase in temperature and decomposition of the product.

4. The submitters used a conventional slide projector bulb (Radiac DYS/SYV/BHC 120V, 600W) and the checkers employed a General Electric DYS 600W/120V bulb. The

bulb was welded to lead wires contained within porcelain tubing that set properly within the inner well of the photoreactor. The bulb was attached to a Variac at full power (the checkers used a Variac set at 480W). Importantly, the bulb was additionally cooled with a strong flow of air that was introduced two-thirds of the way down in the well by means of Tygon tubing appropriately fastened to the porcelain.

5. TLC was carried out by the checkers on silica gel plates using 50:1 chloroform:methanol as eluent and $KMnO_4$ for visualization. Under these conditions, the R_fs of furfural and the furanone are 0.6 and 0.0, respectively.

6. It is convenient to divide the reaction mixture into three portions (ca. 170 mL each) and to transfer them into three 1-L, round-bottomed flasks to effect concentration separately. In each case, some chloroform was introduced and concentration was carried out at 35-38°C. This process was repeated a second time and the residual oils were combined in a 300-mL beaker, this transfer being effected with the aid of a minimum amount of chloroform.

7. The submitters, starting with 266.3 g (2.77 mol) of furfural, obtained the product as colorless crystals, mp 57-59°C and found that, upon stirring the mother liquors at −78°C and seeding, an additional 70.3 g (total yield of 77%) of the butenolide could be obtained. The checkers noted that their product could be decolorized by charcoal treatment (twice) of a chloroform solution heated at reflux to furnish an analytically pure sample of the furanone with mp 54°C. The product displays the following spectral data: ^{1}H NMR (300 MHz, $CDCl_3$) δ: 5.24 (br s, 1H), 6.23 (dd, J= 1.2, 5.4, 1H), 6.27 (m, 1H), 7.33 (dd, J= 1.2, 5.4, 1H); ^{13}C NMR (75 MHz, $CDCl_3$) δ: 99.4, 124.7, 152.5, 172.0; IR (NaCl) cm^{-1}: 3363, 3114, 2929, 2739, 1792, 1756, 1611, 1583, 1442, 1341, 1283, 1183, 1131, 1083, 996; Anal. Calcd for $C_4H_4O_3$: C, 48.01; H, 4.03. Found: C, 48.11; H, 4.15.

8. The submitters purchased d-menthol from TCI and the checkers obtained this compound (99%, 98% ee) from Aldrich Chemical Company, Inc. Menthol was used as received. D-(+)-camphorsulfonic acid (99%) was purchased from Avocado and used as received. Benzene was purchased from EM Science and used as received.

9. The submitters observed that the two liquid phases that are present when refluxing begins form a homogeneous solution within a few minutes.

10. The theoretical amount of water is 5.9 mL. TLC analysis was conducted by the checkers on silica gel plates using 50:1 chloroform:methanol as eluent and iodine and $KMnO_4$ for visualization. Under these conditions, the R_fs of the product ($KMnO_4$) and menthol (I_2) are 0.59 and 0.46, respectively.

11. This mixture solidifies spontaneously on standing. ^1H NMR analysis (in $CDCl_3$ solution) clearly shows the α- (δ 3.53) and β-diastereomers (δ 3.66) to be present in equal amounts.

12. In order to maximize the purity of the product, rapid cooling should be avoided. The submitters allowed the solution to cool slowly to room temperature in an open beaker in the hood, followed by storage in a refrigerator at +5°C and ultimately at –10°C.

13. The product exhibits the following spectral data: ^1H NMR (300 MHz, $CDCl_3$) δ: 0.79 (d, J= 7.0, 3H), 0.84 (d, J= 7.0 Hz, 3H), 0.94 (d, J= 6.5, 3H), 0.96-1.50 (m, 3H), 1.61-1.72 (m, 2H), 2.08-2.17 (m, 2H), 3.66 (dt, J= 4.5, 10.8, 1H), 6.09 (t, J= 1.2, 1H), 6.20 (dd, J= 1.2, 5.9, 1H), 7.17 (dd, J= 1.2, 5.9, 1H); ^{13}C NMR (75 MHz, $CDCl_3$) δ: 15.9, 21.1, 22.4, 23.3, 25.5, 31.7, 34.4, 40.5, 47.9, 79.3, 100.7, 125.0, 151.1, 171.0; IR (NaCl) cm^{-1}: 2930, 1795, 1752, 1451, 1387, 1347, 1312, 1173, 1132, 1016, 923, 825; Anal. Calcd for $C_{14}H_{22}O_3$: C, 70.56; H, 9.30. Found: C, 70.52: H, 9.16. The submitters found $[\alpha]_D^{20}$+136.9 (EtOH, c 8.2) and the checkers observed $[\alpha]_D^{20}$+135.12 ($CHCl_3$, c 8.2).

14. The submitters obtained 14.2 g of product in Step C. Based on the amount of furanone theoretically present in the mother liquors subjected to Step C, this represents a 35% yield of the desired diastereomer in the epimerization step.

Waste Disposal Information

All toxic materials were disposed of in accordance with "Prudent Practices in the Laboratory"; National Academy Press; Washington, DC, 1995.

3. Discussion

The synthetic utility of (5S)- and (5R)-menthyloxy-2(5H)-furanones has been extensively explored and is well documented. The ready availability of these enantiomerically pure butenolides via the singlet photooxygenation of furfural[2] and their wide range of reactivity contribute to their popularity as chiral building blocks. The present procedure for the preparation of the title compound and the useful epimerization step are modifications of methods originally reported by Feringa and co-workers.[3]

The double bond of these enantiomerically pure butenolides participates readily in asymmetric Diels-Alder reactions,[4-6] 1,4-conjugate additions,[7-15] [2+2] photochemical reactions,[16-18] 1,3-dipolar cycloadditions,[3,19-20] diastereoselective dihydroxylation,[21] and tandem double Michael addition/intramolecular nucleophilic substitution processes.[22,23] As a consequence, they have played a key strategic role in the total syntheses of such structurally varied enantiopure targets as dibenzylbutyrolactone[24-27] and aryltetralin lignans,[28] podophyllotoxin[29-31] and isomers thereof,[32,33] β-lactams,[34] grandisol,[17] terebic acid,[35] isostegane derivatives,[36] and chiral nitronic esters[37] among others.

The current procedure provides material of high enantiomeric purity (>98% ee) starting with inexpensive and easily handled reagents. The two steps proceed in reasonable overall yield and require no chromatographic separation.

1. Evans Chemical Laboratories, The Ohio State University, Columbus, OH 43210.
2. (a) Doerr, I. L.; Willette, R. E. *J. Org. Chem.* **1973**, *38*, 3878; (b) Yuste, F.; Sánchez-Obregón, R. *J. Org. Chem.* **1982**, *47*, 3665.
3. Feringa, B. L.; de Lange, B.; de Jong, J. C. *J. Org. Chem.* **1989**, *54*, 2471.
4. Feringa, B. L.; de Jong, J. C. *J. Org. Chem.* **1988**, *53*, 1125.
5. de Jong, J. C.; Jansen, J. F. G. A.; Feringa, B. L. *Tetrahedron Lett.* **1990**, *31*, 3047.
6. de Jong, J. C.; van Bolhuis, F.; Feringa, B. L. *Tetrahedron: Asymmetry* **1991**, *2*, 1247.
7. Feringa, B. L.; de Lange, B. *Tetrahedron Lett.* **1988**, *29*, 1303.
8. Feringa, B. L.; de Lange, B. *Tetrahedron* **1988**, *44*, 7213.
9. Jansen, J. F. G. A.; Feringa, B. L. *Tetrahedron Lett.* **1989**, *30*, 5481.
10. de Lange, B.; van Bolhuis, F.; Feringa, B. L. *Tetrahedron* **1989**, *45*, 6799.
11. Jansen, J. F. G. A.; Feringa, B. L. *Tetrahedron: Asymmetry* **1990**, *1*, 719.
12. Jansen, J. F. G. A.; Jansen, C.; Feringa, B. L. *Tetrahedron: Asymmetry* **1991**, *2*, 109.
13. Jansen, J. F. G. A.; Feringa, B. L. *Synth. Commun.* **1992**, *22*, 1367.
14. Kang, F.-A.; Yin, H.-Y.; Yin, C.-L. *Chin. Chem. Lett.* **1997**, *8*, 365.
15. Kang, F.-A.; Yu, Z.-Q.; Yin, H.-Y.; Yin, C.-L. *Tetrahedron: Asymmetry* **1997**, *8*, 3591.
16. Hoffmann, N.; Scharf, H.-D.; Runsink J. *Tetrahedron Lett.* **1989**, *30*, 2637.
17. Hoffmann, N.; Scharf, H.-D. *Liebigs Ann. Chem.* **1991**, *12*, 1273.
18. Bertrand, S.; Hoffmann, N.; Pete, J.-P. *Tetrahedron* **1998**, *54*, 4873.
19. de Lange, B.; Feringa, B. L. *Tetrahedron Lett.* **1988**, *29*, 5317.
20. Rispens, M. T.; Keller, E.; de Lange, B.; Zijlstra, R. W. J.; Feringa, B. L. *Tetrahedron: Asymmetry* **1994**, *5*, 607.
21. Sundermann, B.; Scharf, H.-D. *Tetrahedron: Asymmetry* **1996**, *7*, 1995.
22. Huang, H.; Chen, Q. *Tetrahedron: Asymmetry* **1998**, *9*, 4103.

23. Huang, H.; Chen, Q. *Tetrahedron: Asymmetry* **1999**, *10*, 1295.

24. Pelter, A.; Ward, R. S.; Jones, D. M.; Maddocks, P. *Tetrahedron: Asymmetry* **1990**, *1*, 857.

25. Pelter, A.; Ward, R. S.; Jones, D. M.; Maddocks, P. *Tetrahedron: Asymmetry* **1992**, *3*, 239.

26. Pelter, A.; Ward, R. S.; Jones, D. M.; Maddocks, P. *J. Chem. Soc., Perkin Trans. 1* **1993**, 2631.

27. van Oeveren, A.; Jansen, J. F. G. A.; Feringa, B. L. *J. Org. Chem.* **1994**, *59*, 5999.

28. Pelter, A.; Ward, R. S.; Jones, D. M.; Maddocks, P. *J. Chem. Soc., Perkin Trans. 1* **1993**, 2621.

29. van Speybroeck, R.; Guo, H.; Van der Eycken, J.; Vandewalle, M. *Tetrahedron* **1991**, *47*, 4675.

30. Bush, E. J.; Jones, D. W. *J. Chem. Soc., Chem. Commun.* **1993**, 1200.

31. Bush, E. J.; Jones, D. W. *J. Chem. Soc., Perkin Trans. 1* **1996**, 151.

32. Pelter, A.; Ward, R. S.; Storer, N. P. *Tetrahedron* **1994**, *50*, 10829.

33. Pelter, A.; Ward, R. S.; Li, Q.; Pis, J. *Tetrahedron: Asymmetry* **1994**, *5*, 909.

34. Lubben, M.; Feringa, B. L. *Tetrahedron: Asymmetry* **1991**, *2*, 775.

35. Hoffmann, N. *Tetrahedron: Asymmetry* **1994**, *5*, 879.

36. Pelter, A.; Ward, R. S.; Abd-el-Ghani, A. *J. Chem. Soc., Perkin Trans. 1* **1996**, 1353.

37. Kang, F.-A.; Yin, C.-L.; She, S.-W. *J. Org. Chem.* **1996**, *61*, 5523.

Appendix

Chemical Abstracts Nomenclature (Collective Index Number); (Registry Number)

2(5H)-(d-Menthyloxy)-2(5H)-furanone: 2(5H)-Furanone, 5-[[(1S,2R,5S)-5-methyl-2-(1-methyethyl)cyclohexyl]oxy]-, (5S)- (9); (122079-41-8)

2-Hydroxy-2(5H)-furanone: 2(5H)-Furanone, 5-hydroxy- (8, 9); (14032-66-7)

Furfural: 2-Furancarboxaldehyde (9); (98-01-1)

Rose Bengal (9); (11121-48-5)

D-(+)-Camphorsulfonic acid: Bicyclo[2.2.1]heptane-1-methanesulfonic acid, 7,7-dimethyl-2-oxo-, (1S,4R)- (9); (3144-16-9)

SYNTHESIS OF INDOLES BY PALLADIUM-CATALYZED REDUCTIVE N-HETEROANNULATION OF 2-NITROSTYRENES: METHYL INDOLE-4-CARBOXYLATE

(1H-Indole-4-carboxylic acid, methyl ester)

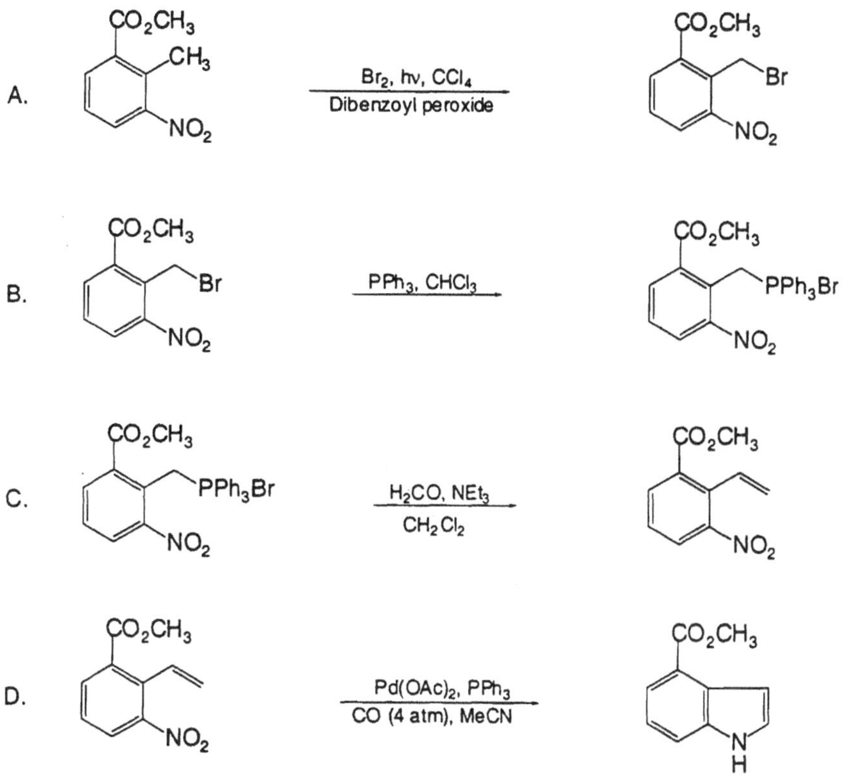

A. Br_2, hν, CCl_4 / Dibenzoyl peroxide

B. PPh_3, $CHCl_3$

C. H_2CO, NEt_3 / CH_2Cl_2

D. $Pd(OAc)_2$, PPh_3 / CO (4 atm), MeCN

Submitted by Björn C. Söderberg,[1] James A. Shriver, and Jeffery M. Wallace.

Checked by J. David Warren and Louis S. Hegedus.

1. Procedure

A. *Methyl 2-bromomethyl-3-nitrobenzoate.* To a 250-mL, two-necked, round-bottomed flask, equipped with a condenser and addition funnel, is added methyl 2-methyl-3-nitrobenzoate (19.1 g, 97.9 mmol, Note 1), dibenzoyl peroxide (1.21 g, 5.00 mmol, Note 2), and 100 mL of carbon tetrachloride (Note 3). The mixture is heated to reflux, and a clear, pale yellow solution is formed. A solution of bromine (16.1 g, 100.6 mmol) (Note 2) in 20 mL of carbon tetrachloride is added over 10 min to the boiling solution under irradiation using a 100-W flood lamp. The reaction mixture is heated and irradiated for 24 hr. The resulting orange solution is allowed to cool to ambient temperature; 50 mL of dichloromethane is added, and the solution is washed with three 50-mL portions of saturated aqueous sodium bicarbonate. The organic phase is dried ($MgSO_4$), filtered, and the solvents are removed at water aspirator pressure on a rotary evaporator affording 25.90 g (96.5%) of methyl 2-bromomethyl-3-nitrobenzoate as pale yellow crystals. The material is used in the next step without further purification (Note 4).

B. *(2-Carbomethoxy-6-nitrobenzyl)triphenylphosphonium bromide.* To a 500-mL, round-bottomed flask is added methyl 2-bromomethyl-3-nitrobenzoate (25.90 g, 94.5 mmol) and 150 mL of chloroform (Note 5). Triphenylphosphine (28.44 g, 108.4 mmol) (Note 2) is added in one portion and the resulting yellow solution is heated to reflux for 1.5 hr. After cooling to ambient temperature, the orange solution is poured into 400 mL of anhydrous diethyl ether (Note 6) with vigorous stirring to precipitate the Wittig salt. The slurry is cooled in a freezer (-20°C, 1 hr). The white solids are collected by filtration, washed with 4 x 100 mL of anhydrous diethyl ether, and dried under high-vacuum to give crude (2-carbomethoxy-6-nitrobenzyl)triphenylphosphonium bromide. A ^{1}H NMR spectrum indicated that the product contained some triphenylphosphine, thus the salt was washed again with diethyl ether (2 x 100 mL). After drying, a considerably purer product (50.91 g) was obtained. The salt is used in the next step without further purification (Note 7).

C. *Methyl 2-ethenyl-3-nitrobenzoate.* Paraformaldehyde (30 g) (Note 8) is placed in a 250-mL, two-necked, round-bottomed flask. An argon inlet is connected to one of the necks. The other neck is connected to a 1-L, two-necked, round-bottomed flask via Tygon tubing (2 cm id) and two male 24/40 joints (Note 9). In the 1-L flask, 2-carbomethoxy-6-nitrobenzyl)triphenylphosphonium bromide (50.91 g) is dissolved in 500 mL of dichloromethane. A glass tube is attached to the Tygon tubing such that the end extends several cm into the solution. Triethylamine (39.13 g, 53.8 mL, 386.7 mmol) is added to the 1-L flask, resulting in the immediate formation of a deep blue/purple solution; a condenser is attached to the second neck of the flask. Argon is then passed over the paraformaldehyde so that a steady stream of gas is constantly bubbling through the purple ylide solution. The paraformaldehyde is heated to 160°C (oil bath) in order to generate formaldehyde, which flows through the solution (Note 10). Upon completion of the addition of formaldehyde, the purple color of the ylide slowly changes from deep purple to brown over a period of 1-2 hr, indicating completion of the reaction. The resulting solution is poured into 500 mL of hexanes, forming a precipitate, which is removed by filtration. The precipitate is washed with 100 mL of hexanes and the combined filtrate and wash are concentrated at water aspirator pressure on a rotary evaporator. The crude yellow-brown solid is dissolved in 100 mL of dichloromethane; 30 g of silica gel is added and the mixture is concentrated again. The resulting powder applied to a pre-packed silica gel column (40 x 4 cm) (Note 11), and eluted with 7:3 hexanes:ethyl acetate to give methyl 2-ethenyl-3-nitrobenzoate (15.87 g, 76.6 mmol, 81%, Notes 12-14) as pale yellow crystals.

D. *Caution! Due to the toxicity of carbon monoxide and the risk of an explosion in handling pressurized glassware, this transformation should be carried out in a well vented fume hood with a blast shield and the hood sash down. Users should exercise appropriate caution at all times.*

Methyl indole-4-carboxylate. To a threaded glass, 200-mL reaction vessel, equipped with a Teflon stirring bar, is added methyl 2-ethenyl-3-nitrobenzoate (10.35 g,

50.0 mmol), triphenylphosphine (3.23 g, 12.3 mmol) and 100 mL of acetonitrile (Note 15). The mixture is stirred for 10 min to dissolve the reagents. Palladium acetate (0.673 g, 3.00 mmol) (Note 16) is added; a yellow precipitate is immediately formed. The tube is attached to a pressure head (Note 17) and the solution is saturated with carbon monoxide (four cycles to 59 psi of CO) (Note 18). The reaction mixture is heated to 90°C (oil bath temperature) under CO (59 psi) for 50 hr. A red-brown solution forms after heating for a few minutes. In order to remove carbon dioxide that is formed and to monitor the progress of the reaction, the vessel is removed from the oil bath for 15 min, carefully vented, and repressurized with CO every 10-12 hr (Note 19). After 50 hr, the reaction mixture is cooled and concentrated using a rotary evaporator at water aspirator pressure, giving a dark brown/black oil. The crude product is purified by chromatography on a pre-packed 40 x 4 cm silica column (Note 11) using 7:3 hexanes:CH_2Cl_2 (0.80 L), then 1:1 hexanes:CH_2Cl_2 (2.80 L) as eluents affording methyl indole-4-carboxylate (7.95 g, 44.75 mmol, 91%) as a pale yellow solid (Note 20).

2. Notes

1. Methyl 2-methyl-3-nitrobenzoate was used as received from Aldrich Chemical Company, Inc., or prepared from 2-methyl-3-nitrobenzoic acid in >97% yield according to the literature procedure.[2]

2. Dibenzoyl peroxide, bromine, and triphenylphosphine were used as received from Aldrich Chemical Company, Inc.

3. Carbon tetrachloride (anhydrous) was used as received from Aldrich Chemical Company, Inc.

4. The checkers did not observe complete conversion to the bromide. Analytical data: mp 63-66°C (uncorrected). ^{1}H NMR (300 MHz, $CDCl_3$) δ: 3.97 (s, 3 H), 5.13 (s, 2 H), 7.52 (t, J = 8.1, 1 H), 7.93 (dd, J = 8.1, 1.2, 1 H), 8.08 (dd, , J = 7.7, 1.2, 1 H); ^{13}C

NMR (75 MHz, $CDCl_3$) δ: 23.2, 53.4, 128.0, 129.4, 132.5, 132.8, 134.9, 150.7, 166.0; IR (neat) cm^{-1}: 1726, 1532, 1270. Anal. Calcd for $C_9H_8BrNO_4$: C, 39.44; H, 2.94. Found: C, 39.33; H, 2.94.

5. Chloroform (anhydrous) was used as received from Aldrich Chemical Company, Inc.

6. Diethyl ether (anhydrous) was used as received from Fisher Scientific.

7. Analytical data: mp 217-229°C (sealed capillary, uncorrected); ^1H NMR (300 MHz, $CDCl_3$) δ: 3.67 (s, 3 H), 5.69 (d, J_{PH} = 14.6, 2 H), 7.55-7.82 (overlapping multiplets, 16 H), 7.95 (d, J = 8.1, 1 H), 8.09 (d, J = 7.9, 1 H); ^{13}C NMR (75 MHz, $CDCl_3$) δ: 26.3 (d, J_{CP} = 51.7), 53.5, 117.7 (d, J_{CP} = 86.4), 124.8 (d, J_{CP} = 8.4), 129.1 (d, J_{CP} = 2.8), 130.0 (d, J_{CP} = 12.5), 130.8, 133.8 (d, J_{CP} = 9.8), 135.2 (d, J_{CP} = 2.9), 135.7, 150.9, 165.5; IR (neat) cm^{-1}: 1717, 1532, 1437, 1275, 1108.

8. Paraformaldehyde was used as received from Fisher Scientific.

9. Formaldehyde is generated according to the procedure of Smith, A. B., III; Branca, S. J.; Guaciaro, M. A.; Wovkulich, P. M. *Org. Synth., Coll. Vol. VII* **1990**, 271; see Figure 1, p. 272.

10. The glass inlet tube has a tendency to clog. This problem can be remedied by lifting the tube out of the solution, discontinuing heating, and removing the obstruction. It is important to closely follow the pyrolysis. If insufficient heat is supplied, the ylide solution may be sucked into the flask containing paraformaldehyde. If too much heat is supplied, the joints may separate, usually resulting in a small flame at the joint.

11. Silica gel (200-400 mesh) from Natland International Corp. was used.

12. The yield is based on methyl-2-bromomethyl-3-nitrobenzoate.

13. A small amount (≤1.5%) of hydrolysis product, methyl 2-methyl-3-nitrobenzoate, was observed at times.

14. Analytical data: mp 39-40°C (uncorrected); ^1H NMR (300 MHz, $CDCl_3$) δ: 3.89 (s, 3 H), 5.23 (dd, J = 17.6, 0.8, 1 H), 5.43 (dd, J = 11.3, 0.8, 1 H), 7.18 (dd,

J = 17.6, 11.5, 1 H), 7.47 (t, J = 7.9, 1 H), 7.88 (dd, J = 8.3, 1, 1 H), 7.98 (dd, J = 7.7, 1.2, 1 H); ^{13}C NMR (75 MHz, $CDCl_3$) δ: 52.9, 119.7, 126.5, 128.1, 132.1, 132.2, 133.2, 134.2, 150.4, 166.8; IR (neat) cm^{-1}: 1730, 1531, 1292, 1264, 1124, 707. Anal. Calcd for $C_{10}H_9NO_4$: C, 57.97; H, 4.38. Found: C, 57.72; H, 4.47.

15. Acetonitrile was distilled from calcium hydride prior to use.

16. Palladium acetate was used as received from Pressure Chemical Co.

17. The pressure head is assembled as follows using, if possible, stainless steel parts: A tee, equipped with a pressure relief valve (preset to open at 120 psi) and a pressure gauge (0-400 psi), is connected to a cross via a nipple. A ball valve (for pressure release and sample removal) is attached to the top of the cross, a valve (for introduction of CO) on one side of the cross, and a nipple connected to a Swagelok Teflon adapter attached to the bottom of the cross. An autoclave or a metal reaction vessel can be substituted for this reaction assembly.

18. UHP-grade carbon monoxide is used as purchased from Matheson Gas Products, Inc.

19. The reaction is monitored by thin layer chromatography on 60 Å silica gel (R_f = 0.15; 4:6 hexanes:CH_2Cl_2).

20. After ca. 500 mL of solvent, 20-mL fractions are collected and analyzed by thin layer chromatography. The plates are visualized at 254 nm, and the product appears as a bright fluorescent blue spot. Test tubes containing the product are combined into two major fractions the first (1.173 g) contaminated by a small amount of an unknown impurity (higher R_f) and the second (6.663 g) as pure methyl indole-4-carboxylate. The ^1H NMR spectra of the two fractions are identical; however; a slight melting point depression is observed for the first fraction. Uncorrected melting points: first fraction 64-66°C; second fraction 68-69°C. Note: A range of melting points has been reported for this compound: Lit. mp 63°C;[3] 64-65°C;[4] 67-69°C;[2] 69-71°C.[5] ^1H NMR (300 MHz, $CDCl_3$) δ: 3.99 (s, 3 H), overlapping 7.19 (br d, J = 3.0, 1 H), 7.23 (t, J = 7.7, 1 H), 7.34 (t, J = 3.0, 1 H), 7.59

(d, J = 8.1, 1 H), 7.93 (d, J = 7.5, 1 H), 8.48 (br s, 1 H); ^{13}C NMR (75 MHz, $CDCl_3$) δ: 52.2, 103.9, 116.5, 121.3, 121.6, 123.6, 126.8, 127.6, 136.9, 168.4; IR (neat) cm^{-1}: 3355, 1699, 1276.

Waste Disposal Information

All toxic materials were disposed of in accordance with "Prudent Practices in the Laboratory"; National Academy Press; Washington, DC, 1995.

3. Discussion

The procedure described here illustrates an efficient and relatively mild synthesis of methyl indole-4-carboxylate in 72% overall yield starting from commercially available materials. The synthetic sequence compares favorably to the previously described Batcho-Leimgruber indole syntheses of this substance.[2,3] In general, the palladium/phosphine-catalyzed N-heteroannulation of 2-nitrostyrenes offers a very flexible entry to functionalized indoles.[6] A few examples of this reaction, performed on a 1-2 mmol scale, are shown in the Table below. A number of palladium reagents, both palladium(0) and palladium(II), can be used as precatalyst for the reaction. For example, palladium diacetate, bis(acetonitrile)palladium dichloride, bis(dibenzylidenacetone)palladium and palladium on carbon (10% Pd) have all been shown to produce indoles in the presence of carbon monoxide and a catalytic amount of phosphine.[6a] Perhaps the most important feature of the reaction is its compatibility with other functional groups. For example, functionalities such as bromides, triflates, alcohols, ethers, esters, ketones, nitriles, and additional nitro groups all remain unaffected in this reaction. In contrast to the Batcho-Leimgruber route, indoles substituted in the 2- and/or 3-position can readily be prepared from 2-nitrostyrenes. More complex, fused indoles, can be obtained by annulation of bicyclic nitrostyrene derivatives (entries 7-8).

1. Department of Chemistry, West Virginia University, Morgantown, WV 26506.

2. Hoffmann-La Roche and Co., British Patent 1 276 966; **1972**; Batcho, A. D. and Leimgruber, W. German Patent 2 057 840; **1971**.

3. Ponticello, G. S.; Baldwin, J. J. *J. Org. Chem.* **1979**, *44*, 4003; Kozikowski, A. P.; Ishida, H.; Chen, Y.-Y. *J. Org. Chem.* **1980**, *45*, 3350.

4. Watanabe, T.; Hamaguchi, F.; Ohki, S. *Chem. Pharm. Bull.* **1972**, *20*, 2123.

5. Aldrich Handbook of Fine Chemicals and Laboratory Equipment, **2000-2001**, 1126.

6. (a) Söderberg, B. C.; Rector, S. R.; O'Neil, S. N. *Tetrahedron Lett.* **1999**, *40*, 3657; (b) Söderberg, B. C.; Shriver, J. A. *J. Org. Chem.* **1997**, *62*, 5838 and references therein. (c) Söderberg, B. C.; Chisnell, A. C.; O'Neil, S. N.; Shriver, J. A. *J. Org. Chem.* **1999**, *64*, 9731.

Appendix

Chemical Abstracts Nomenclature (Collective Index Number); (Registry Number)

Methyl indole-4-carboxylate: 1H-Indole-4-carboxylic acid, methyl ester (9); (39830-66-5)

Methyl 2-methyl-3-nitrobenzoate: Benzoic acid, 2-methyl-3-nitro-, methyl ester (9); (59382-59-1)

Methyl 2-bromomethyl-3-nitrobenzoate: Benzoic acid, 2-bromomethyl-3-nitro-, methyl ester (9); (98475-07-1)

Dibenzoyl peroxide: Peroxide, dibenzoyl (8,9); (94-36-0)

(2-Carbomethoxy-6-nitrobenzyl)triphenylphosphonium bromide: Phosphonium, [[2-(methoxycarbonyl)-6-nitrophenyl]methyl]triphenyl-, bromide; (195992-09-7)

Methyl 2-ethenyl-3-nitrobenzoate: Benzoic acid, 2-ethenyl-3-nitro-, methyl ester; (195992-04-2)

Triphenylphosphine: Phosphine, triphenyl- (8,9); (603-35-0)

Paraformaldehyde (8,9); (30525-89-4)

Triethylamine: Ethanamine, N.N-diethyl- (9); (121-44-8)

Carbon monoxide (8,9); (630-08-0)

Palladium acetate: Acetic acid, palladium(2+) salt (8,9); (3375-31-3)

Table			
Indoles By Reductive N-Heteroannulation of 2-Nitrostyrenes			
Entry	**Styrene**	**Indole**	**Yield (%)**
1	NO_2	H	97
2	OH NO_2	OH H	96
3	NO_2 NO_2	NO_2 H	89
4	CH_2OMe NO_2	CH_2OMe H	76
5	CO_2Me NO_2	CO_2Me H	90
6	MeO NO_2	MeO H	81
7	Ac N NO_2	Ac N H	41
8	O NO_2	O H	63

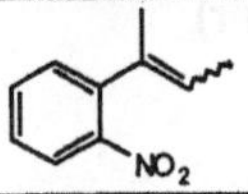

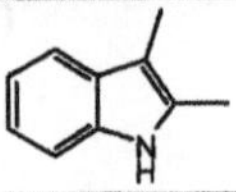

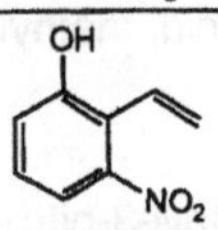

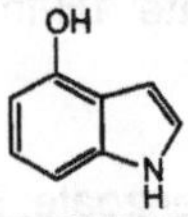

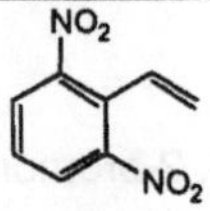

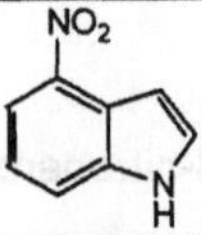

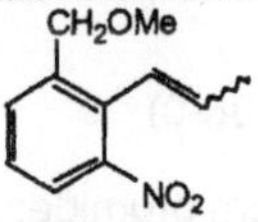

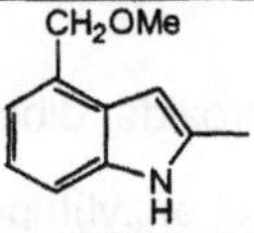

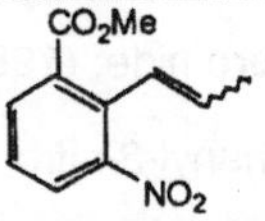

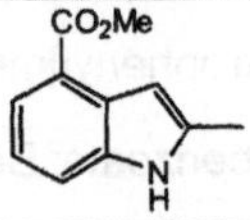

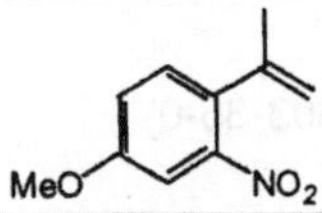

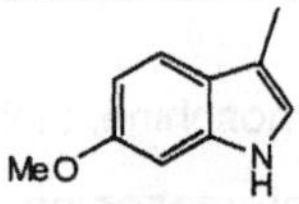

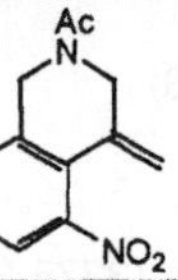

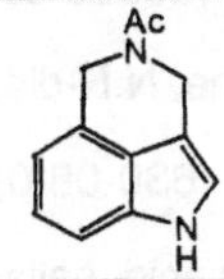

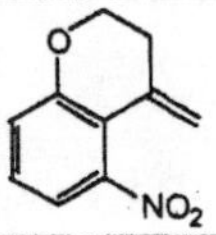

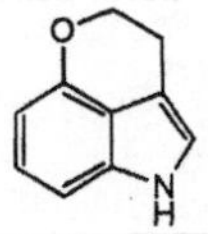

RING-CLOSING METATHESIS SYNTHESIS OF N-Boc-3-PYRROLINE

(1H-Pyrrole-1-carboxylic acid, 2,5-dihydro-, 1,1-dimethylethyl ester)

Boc — N-diallylamine → 1. $(PCy_3)_2Cl_2Ru{=}CHPh$ (cat); 2. $P(CH_2OH)_3/H_2O$ → N-Boc-3-pyrroline

Submitted by Marcelle L. Ferguson,[1] Daniel J. O'Leary,[1] and Robert H. Grubbs.[2]

Checked by Louise M. Stamp and Andrew B. Holmes.

1. Procedure

N-Boc-3-pyrroline.[3] An oven-dried, 3-L, three-necked, round-bottomed flask equipped with an overhead mechanical stirrer, rubber septum with nitrogen inlet, and condenser with a bubbler-sealed outlet, is charged with 732 mg (0.89 mmol, 0.5 mol%) of bis(tricyclohexylphosphine)benzylidine ruthenium dichloride (Note 1) and 445 mL of dry dichloromethane (Note 2). The mixture is stirred under a nitrogen atmosphere to give a burgundy-colored solution. N-Boc-diallylamine (38.3 mL, 178 mmol) (Note 3) is transferred to the flask by syringe. After addition of the N-Boc-diallylamine, ethylene gas is evolved vigorously and the color of the solution changes from burgundy to dark brown. The solution is heated to reflux with stirring for 2.5 hr (Note 4), then cooled to room temperature.

An aqueous methanolic solution of tris(hydroxymethyl)phosphine is prepared (Note 5). This solution is transferred to the flask containing the N-Boc-3-pyrroline, using 5 mL of methanol as a rinse. Triethylamine (0.247 mL, 1.78 mmol) is added to which is vigorously stirred at room temperature overnight under a nitrogen atmosphere (Note 9).

Deionized water (400 mL) is added to the flask and the mixture is stirred vigorously for 30 min. Using a 2-L separatory funnel, the dichloromethane phase is separated from the aqueous phase, which is discarded. The organic phase is then washed with 350 mL of deionized water and with 200 mL of 50% v/v water/saturated brine solution. The dichloromethane phase is dried with 10 g of $MgSO_4$, then filtered through a sintered glass funnel. The filtrate is concentrated under reduced pressure by rotary evaporation (water aspirator) using a bath temperature of ca. 40°C.

The crude product is divided in half and each portion is separately distilled in a Kugelrohr apparatus (50 mL distillation and receiver flasks) (Notes 10, 11) to give a combined yield of 27.2-28.3 g (90-94%) of N-Boc-3-pyrroline as a white, crystalline, low-melting solid (Note 12).

2. Notes

1. Bis(tricyclohexylphosphine)benzylidine ruthenium dichloride was obtained from Strem Chemical Company and used as received.

2. Dichloromethane was distilled over calcium hydride before use. The submitters found that identical reaction yields were obtained using commercial amylene-free HPLC-grade dichloromethane.

3. N-Boc-diallylamine was obtained from Aldrich Chemical Company, Inc. and used as received.

4. The progress of the reaction can be conveniently monitored by gas chromatography. GC analyses were performed on a Hewlett-Packard 6890 series gas chromatograph equipped with a 6890 series mass selective detector and an HP-5MS 30-m x 0.25-mm x 0.25-µm column under the following conditions: injector temp 180°C; detector temp 150°C; oven temp 70°C, 2 min; ramp 5.75°C/min; final temp 150°C;

150°C; helium gas flow 20.0 mL/min; t_R = 9.73-9.75 min (the submitters observed t_R 9.71 min). Complete conversion was usually observed after 2 hr of reflux.

5. Preparation of tris(hydroxymethyl)phosphine. A 250-mL, three-necked, round-bottomed flask, equipped with a condenser having a bubbler-sealed outlet, rubber septum with nitrogen inlet, and magnetic stir bar, is charged with 43 mL (74.1-79.1 mmol) of a 1.72-1.84M aqueous solution of tetrakis(hydroxymethyl)phosphonium sulfate (Note 6). The flask is immersed in an oil bath. Methanol (25 mL) is added to the flask (Note 7), and the contents are heated to a gentle reflux under a nitrogen atmosphere. Sodium hydroxide pellets (3.1 g, 76.5 mmol) are added to the flask over the course of 30 min, accompanied by the gradual addition of 40 mL of methanol (Note 8). The mixture is stirred for an additional 10 min, then cooled to room temperature with stirring.

6. Tetrakis(hydroxymethyl)phosphonium sulfate was obtained as a 70-75% w/v (ca. 1.72-1.84M) aqueous solution from Fluka and was not purified before use. The submitters used 30.5 mL (76.5 mmol) of a 2.56M aqueous solution (11% active phosphorus by weight) obtained from Cytec, Inc.

7. Methanol was added to the tetrakis(hydroxymethyl)phosphonium sulfate by syringe through the rubber septum and this resulted in a cloudy, white solution.

8. A white precipitate forms on the bottom of the flask as NaOH is added over the 30 min period, causing the solution to become viscous. Additional methanol is added to enable stirring to be maintained.

9. The brown color of the N-Boc-3-pyrroline solution begins to fade immediately upon addition of tris(hydroxymethyl)phosphine. Approximately 45 min later, a viscous, brown substance is observed to adhere to the walls of the flask. This initially insoluble material dissolves after stirring overnight. A pale yellow solution results.

10. The submitters used a Kugelrohr distillation apparatus fitted with 24/40 ground glass joints and the N-Boc-3-pyrroline was distilled from a 500-mL, round-bottomed flask directly into a 250-mL receiving flask.

11. The Kugelrohr oven was heated from room temperature to 80°C over the course of 45 min. The submitters noted that the pressure was 280 mtorr as the residual solvent was removed. As the oven temperature approached 50°C, an acetone and dry ice bath was placed under the collection bulb. The submitters noted that the system pressure then dropped to 70 mtorr and the product distilled as a clear and colorless liquid that rapidly crystallized. The checkers observed similar results. The distillation was terminated when the distilling flask was nearly empty.

12. The submitters obtained 28.71 g (96%) of N-Boc-3-pyrroline. The analytical properties are as follows: mp 36-38°C; ^{1}H NMR (400 MHz, $CDCl_3$) δ: 1.46 (s, 9 H), 4.10 (s, 4 H), 5.75 (s, 2 H); ^{13}C NMR (100 MHz, $CDCl_3$) δ: 28.5, 53.1, 79.2, 125.8, 154.3; IR ($CHCl_3$) cm^{-1} 2937, 1680, 1623, 1522, 1477, 1405, 1208, 1124, 995, 950, 871, 742, 624; MS (EI) m/z (rel intensity) 169 (20, M^+), 114 (30), 96 (45), 68 (100); HRMS (ES^+) m/z 192.1004 ($[M + Na]^+$, calcd. for $C_9H_{15}NO_2Na$ 192.1000). Anal. Calcd. for $C_9H_{15}NO_2$: C, 63.9; H, 8.9; N, 8.3. Found C, 63.3; H, 8.8; N, 8.1.

Waste Disposal Information

All toxic materials were disposed of in accordance with "Prudent Practices in the Laboratory"; National Academy Press; Washington, DC, 1995.

3. Discussion

The preparation of 3-pyrroline was recently reported in *Organic Syntheses*.[4] The protected form, N-Boc-3-pyrroline, has been used effectively in Heck arylations for the preparation of various 4-aryl endocyclic enecarbamates.[5] It has also been used in the preparation of regioisomeric 3-hydroxyisoxazolinyl prolines which are medicinally active.[6] The catalytic asymmetric hydroformylation of N-Boc-pyrroline has been investigated.[7] N-Boc-pyrroline has also been utilized in the synthesis of various antibacterial compounds.[8,9,10]

A procedure to remove colored ruthenium impurities using lead tetraacetate oxidation followed by filtration through a silica gel plug has been described.[11]

1. Department of Chemistry, Pomona College, Claremont, CA 91711.
2. Division of Chemistry and Chemical Engineering, California Institute of Technology, Pasadena, CA 91125.
3. For the original reports from which this procedure was adapted, see: (a) Fu, G. C.; Nguyen, S. T.; Grubbs, R. H. *J. Am. Chem. Soc.* **1993**, *115*, 9856; (b) Maynard, H. D.; Grubbs, R. H. *Tetrahedron Lett.* **1999**, *40*, 4137.
4. Meyers, A. I.; Warmus, J. S.; Dilley, G. J. *Org. Synth., Coll. Vol. IX*, **1998**, 666.
5. Carpes, M. J. S.; Correia, C. R. D. *Synlett* **2000**, *7*, 1037.
6. Conti, P.; Dallanoce, C.; De Amici, M.; De Micheli, C.; Fruttero, R. *Tetrahedron* **1999**, *55*, 5623.
7. Horiuchi, T.; Ohta, T.; Shirakawa, E.; Nozaki, K.; Takaya, H. *J. Org. Chem.* **1997**, *62*, 4285.
8. Hong, C. Y.; Kim, Y. K.; Lee, Y. H.; Kwak, J. H. *Bioorg. Med. Chem. Lett.* **1998**, *8*, 221.
9. Lee, J. W.; Son, H. J.; Lee, K. S.; Yu, Y. H.; Yoon, G. J. *Yakhak Hoechi* **1994**, *38*, 677. *Chem. Abstr.* **1994**, *122*, 235095.
10. Okada, T.; Sato, H.; Tsuji, T.; Tsushima, T.; Nakai, H.; Yoshida, T.; Matsuura, S. *Chem. Pharm. Bull.* **1993**, *41*, 132.

11. Paquette, L. A.; Schloss, J. D.; Efremov, I.; Fabris, F.; Gallou, F.; Mendez-Andino, J.; Yang, J. *Org. Lett.* **2000**, *2*, 1259.

12. Crimmins, M. T.; King, B. W. *J. Org. Chem.* **1996**, *61*, 4192.

13. Zuercher, W. J.; Scholl, M.; Grubbs, R. H. *J. Org. Chem.* **1998**, *63*, 4291.

14. Zuercher, W. J.; Hashimoto, M.; Grubbs, R. H. *J. Am. Chem. Soc.* **1996**, *118*, 6634.

15. Barrett, A. G. M.; Baugh, S. P. D.; Gibson, V. C.; Giles, M. R.; Marshall, E. L.; Procopiou, P. A. *Chem. Commun.* **1997**, 155.

16. Barrett, A. G. M.; Baugh, S. P. D.; Gibson, V. C.; Giles, M. R.; Marshall, E. L.; Procopiou, P. A. *Chem. Commun.* **1996**, 2231.

17. Coates, G. W.; Grubbs, R. H. *J. Am. Chem. Soc.* **1996**, *118*, 229.

18. Dyatkin, A. B. *Tetrahedron Lett.* **1997**, *38*, 2065.

19. Rutjes, F. P. J. T.; Schoemaker, H. E. *Tetrahedron Lett.* **1997**, *38*, 677.

20. Schuster, M.; Pernerstorfer, J.; Blechert, S. *Angew. Chem. Int. Ed. Eng.* **1996**, *35*, 1979.

21. Kinoshita, A.; Mori, M. J. *Org. Chem.* **1996**, *61*, 8356.

Appendix

Chemical Abstracts Nomenclature (Collective Index Number); (Registry Number)

N-Boc-3-pyrroline: 1H-Pyrrole-1-carboxylic acid, 2,5-dihydro-, 1,1-dimethylethyl ester (9); (73286-70-1)

Bis(tricyclohexylphosphine)benzylidine ruthenium (IV) dichloride: Ruthenium, dichloro(phenylmethylene)bis(tricyclohexylphosphine)-, (SP-5-31)- (9); (172222-30-9)

Boc-diallylamine: Carbamic acid, di-2-propenyl-, 1,1-dimethylethyl ester (9); (151259-38-0)

Tetrakis(hydroxymethyl)phosphonium sulfate ("Pyroset-TKOW"): Phosphonium, tetrakis(hydroxymethyl)-, sulfate (2:1) (9); (55566-30-8)

Tris(hydroxymethyl)phosphine: Methanol, phosphinidynetris- (9); (2767-80-8)

Triethylamine: Ethanamine, N,N-diethyl- (9); (121-44-8)

Table 1. Examples of RCM using Grubb's Catalyst, 1

Entry	Substrate	Conditions (yield)	Product	Reference
1	O O OH O N Ph H **3**	1 mol% 1, 30 min, 25°C, CH_2Cl_2 (97%)	O O OH O N Ph H **4**	12
2	O O O **5**	8 mol% 1, 4 hr, 45°C, C_6H_6, (60%)	O O O **6**	13
3	O O **7**	5 mol% 1, 2 hr, 60°C, C_6H_6, (90%)	O O **8**	14
4	OTBDMS H N O **9**	5 mol% 1, 6 hr, 25°C, CH_2CL_2, (78%)	OTBDMS H N O **10**	15
5	O H_2N O O **11**	5 mol% 1, 1 hr, 25°C, CH_2Cl_2, (83%)	O O O NH_2 **12**	16
6	**13**	2 mol% 1, 21 hr, 35°C, CH_2Cl_2, (>97%)	**14**	17
7	Bn O N–Ph N O **15**	10 mol% 1, overnight, reflux, CH_2Cl_2, (>95%)	Bn O N–Ph N O **16**	18
8	H N CO_2Me H Boc **17**	5 mol% 1, 30 min, reflux, CH_2Cl_2, (93%)	H N CO_2Me H Boc **18**	19
9	P O CF_3CON Ph **19**	13 mol% 1, 12 hr, reflux, CH_2Cl_2, (70%) (tritylpolystyrol resin)	P O CF_3CON Ph **20**	20
10	MeO_2C H N O **21**	4 mol% 1, 5 hr, 25°C, CH_2Cl_2, (87%)	MeO_2C H N O **22**	21

THE CATALYTIC INTRAMOLECULAR PAUSON-KHAND REACTION: 2,3,3α,4-TETRAHYDRO-2-[(4-METHYLBENZENE) SULFONYL]CYCLOPENTA[C]PYRROL-5(1H)-ONE

(Cyclopenta[b]pyrrol-5(1H)-one, 2,3,3a,4-tetrahydro-1-[(4-methylphenyl)sulfonyl]-)

A. H_2N (1) → TsCl, pyridine, 10°C → TsNH (2)

B. TsNH (2) → K_2CO_3, acetone, C_3H_3Br, reflux → TsN (3)

C. HO, Me, Me, H (4) → $Co_2(CO)_8$, Et_2O → $(OC)_3Co$, HO, Me, Me, $Co(CO)_3$ (5)

D. $(OC)_3Co$, HO, Me, Me, $Co(CO)_3$ (5) → 1. Et_3SiH, $CycNH_2$; 2. 3,1,2-DME, 67°C → TsN, O (6)

Submitted by Mittun C. Patel,[1] Tom Livinghouse,[2] and Brian L. Pagenkopf.[1]

Checked by Kyung-Hee Kim and Marvin J. Miller.

1. Procedure

A. *N-(2-Propenyl)-4-methylbenzenesulfonamide (2).* A 1-L, two-necked, round-bottomed flask equipped with a magnetic stirring bar, internal thermometer and powder funnel is charged with p-toluenesulfonyl chloride (97.2 g, 0.51 mol) (Note 1). The powder funnel is replaced with a rubber septum connected to a positive pressure of argon and an oil bubbler. The apparatus is flushed with argon and charged with tetrahydrofuran (THF) (400 mL, Note 2) and pyridine (42.9 mL, 0.53 mol). The flask is placed in an ice bath and, after the reaction mixture has cooled to ca. 10°C, allylamine (37.5 mL, 0.50 mol) (Note 3) is added portion-wise by syringe over ca. 40 min (exothermic) maintaining the internal temperature below 15°C. The ice bath is removed and the resulting solution is allowed to warm to room temperature. After 4 hr the rubber septum is removed and the mixture is treated with 75 mL of a 2M aqueous solution of sodium hydroxide. After another 4 hr, the reaction mixture is transferred to a separatory funnel, the organic phase is separated, and the aqueous phase is extracted with two 100-mL portions of ethyl acetate (EtOAc). The combined organic phases are washed with brine (50 mL) and dried with magnesium sulfate ($MgSO_4$) in the presence of activated carbon (4 g). The solution is filtered through a plug of silica gel (diameter: 5 cm; height: 3 cm) and the cake is washed with EtOAc (300 mL). The combined filtrate and washes are concentrated under reduced pressure. The crude product is recrystallized from 275 mL of 30% EtOAc/hexanes to afford 65 g (62%) of allyl tosylamide as a first crop. A second crop of 27 g (26%, 88% in total) is obtained from the mother liquor (Note 4).

B. *N-(2-Propenyl)-N-(2-propynyl)-4-methylbenzenesulfonamide (3).* A 500-mL, single-necked, round-bottomed flask equipped with a Teflon-coated stirring bar (Note 1) is charged with allyl tosylamide (31.7 g, 150 mmol), anhydrous potassium carbonate (K_2CO_3) (24.8 g, 1.2 equiv, 180 mmol) (Note 5), 1-bromo-2-propyne (20.0 mL, 1.2 equiv, 180 mmol), and acetone (300 mL). The flask is equipped with a water-cooled condenser fitted with a rubber septum. The apparatus is flushed with argon introduced through the septum and a positive pressure of argon is maintained with an argon-filled balloon (Note 6). The reaction mixture is heated to reflux with stirring for 24 hr. After complete consumption of starting material, monitored by thin layer chromatography (TLC, Note 7), the reaction mixture is allowed to cool and is concentrated under reduced pressure on a rotary evaporator. The residue is diluted with EtOAc (250 mL) and water (125 mL) and the organic phase is separated. The aqueous phase is extracted with 200 mL of EtOAc and the combined organic phases are washed with brine (50 mL), dried ($MgSO_4$), filtered and concentrated under reduced pressure with a rotary evaporator. The residue is recrystallized from 2.5 mL of ether and 250 mL of 20% EtOAc/hexanes to afford ca. 30 g (80%) of **3** as nearly colorless crystals. A second crop totaling 5.9 g (16%, 96% in total) is also obtained (Note 8).

C. *Hexacarbonyl[μ-[(3,4-η:3,4-η)-2-methyl-3-butyn-2-ol]]dicobalt* (**5**). A single-necked, 50-mL, round-bottomed flask (Note 9), equipped with a Teflon-coated stirring bar, is charged with dicobalt octacarbonyl ($Co_2(CO)_8$) (1.7 g, 5.0 mmol, 1.0 equiv) (Note 10). After attaching a rubber septum, the flask is flushed with carbon monoxide (CO), then charged with 25 mL of degassed diethyl ether (Note 11). The flask is placed in an ice bath and, after 10 min, a bubbler is connected through the septum, followed immediately by the addition of 2-methyl-3-butyn-2-ol (0.51 mL, 5.3 mmol, 1.05 eq). After removing the ice bath and stirring for 3 hr at ambient temperature, the resultant solution is filtered through a pad of Celite (diameter: 3 cm; height: 1 cm) (Note 12), which is washed with ether until the washes are clear (ca. 75 mL).

The alkyne cobalt complex is purified by silica gel chromatography. To the red solution is added 2 g of silica gel and the mixture is concentrated on a rotary evaporator (Note 13). The silica gel is then placed under vacuum (1 mm) until effervescence ceases (about 10 min). The dried silica gel is placed at the top of a column of silica gel (diameter: 3 cm; height: 9 cm), eluting first with hexane to remove non-polar by-products and subsequently with 15% ether/hexane. The desired product elutes as a red band. Concentration of the appropriate fractions on a rotary evaporator (Note 13) and under high vacuum (0.7 mm for 10 min) affords 1.4 g (79%) of **5** as a fine red powder (Note 14).

D. *2,3,3α,4-Tetrahydro-2-[(4-methylbenzene)sulfonyl]cyclopenta[c]pyrrol-5(1H)-one* (**6**). A 250-mL, two-necked, round-bottomed Schlenk flask, equipped with a thermometer and Teflon-coated stirring bar, is charged with 6.0 g (24.1 mmol) of enyne **3** prepared in Step B (Note 1). A 12-cm, air-cooled condenser is attached to the flask. The top of the condenser is fitted with a rubber septum, which is connected to a CO supply and the apparatus is flushed with CO through the Schlenk valve (Note 15) for several min. After the CO purge is completed, the flask is charged with 120 mL of 1,2-dimethoxyethane (1,2-DME) (Note 16). Meanwhile, another 50-mL, round-bottomed Schlenk flask is charged with the $Co_2(CO)_6$-alkyne complex **5** (445 mg, 1.20 mmol, 5 mol%) (Note 9) and flushed with CO with a CO-filled balloon as described above. To the flask containing the $Co_2(CO)_6$-alkyne complex **5** are added sequentially 22.5 mL of 1,2-DME; cyclohexylamine (0.413 mL, 3.61 mmol, 15 mol%), 3 eq of amine/cobalt dimer) and triethylsilane (Et_3SiH) (2.40 mL of a 0.5M solution in xylenes, 1.20 mmol, 5 mol%). The assembly is heated in an oil bath at 67°C for 15 min (Note 17), then transferred via cannula under an atmosphere of CO to the Schlenk flask containing enyne **3** (Note 18). The resulting mixture is heated at 67°C for 24 hr, when the reaction is usually complete as determined by TLC analysis (Note 19). After complete carbocyclization, the reaction mixture is transferred (in air) to a single-necked flask and concentrated on a rotary evaporator. The dark residue is dissolved in 150 mL of dichloromethane (CH_2Cl_2), washed with 10% aqueous sulfuric acid (2 x 15 mL), water (15 mL), and brine (45 mL) and treated with ca. 4 g of activated carbon and anhydrous sodium sulfate (Na_2SO_4). After 4 hr the solution is filtered through a pad of Celite (diameter: 5 cm; height: 2 cm) (Note 12) and concentrated on a rotary evaporator to afford a brown solid. Purification by flash chromatography with silica gel (Note 20) affords 5.8-6.2 g (86-93% yield) of **6** as a slightly yellow solid. Colorless microcrystalline material may be obtained by recrystallization from a cold (–20°C) solution of 1:1 2-propanol:CH_2Cl_2 (4.6-5.1 g, 69-77%) (Notes 21, 22).

2. Notes

1. All glassware is dried in a 200°C oven overnight prior to use, assembled while hot, and allowed to cool to room temperature under dry argon.

2. Mallinckrodt ChromAR® HLPC grade THF was used as received.

3. Allylamine was purchased from Acros Organics and distilled from calcium sulfate before use.

4. The checkers obtained 80 g (76%). Properties of **2**[10]: mp 58.0-59.0°C (EtOAc/hexanes); ^{1}H NMR ($CDCl_3$) δ: 2.34 (s, 3H), 3.49 (app t, J = 5.95, 2H), 4.96-5.12 (m, 2H), 5.23 (s, 1H), 5.57-5.68 (m, 1H), 7.21 (d, J = 8.2, 2H), 7.69 (d, J = 8.2, 2H); ^{13}C NMR ($CDCl_3$) δ: 21.2, 45.5, 117.2, 126.9, 129.5, 132.9, 136.9, 143.2.

5. Unless otherwise noted, all solvents and reagents were purchased from VWR Scientific or Aldrich Chemical Co., Inc. and used as received. $CDCl_3$ was filtered through basic alumina (Aldrich Chemical Co., Inc.) prior to use.

6. The checkers used a double needle in the septum, one for argon supply and the other attached to an argon-filled balloon.

7. The R_f of the starting material and product is 0.30 and 0.52, respectively (1:1 EtOAc:hexanes).

8. The checkers obtained 33.2 g (89%). Properties of **3**: mp 60.8-61.5°C (EtOAc/hexanes); ^{1}H NMR ($CDCl_3$) δ: 1.96 (t, J = 2.4, 1 H), 2.35 (s, 3H), 3.76 (d, J = 6.5, 2H), 4.02 (d, J = 2.4, 2H), 5.25-5.14 (m, 2H), 5.72-5.59 (m, 1 H), 7.23 (d, J = 8.1, 2H), 7.66 (dd, J = 8.3, 1.5, 2H); ^{13}C NMR ($CDCl_3$) δ: 21.3 (CH_3), 35.7 (CH_2), 48.9 (CH_2), 73.6 (CH_2), 119.7 (CH_2), 127.6 (CH), 129.3 (CH), 131.8 (CH), 136.0 (C), 143.4 (C); IR (thin film) cm^{-1}: 3257, 3049, 2987, 2885, 2117, 1646, 1601, 1425, 1330, 1165.

9. The glassware must be cooled to room temperature before use.

10. *Caution. Dry $Co_2(CO)_8$ can be pyrophoric.* $Co_2(CO)_8$ is toxic and should be handled in a glove box or well-ventilated hood. $Co_2(CO)_8$ was purchased from Strem Chemical Co., Inc. High purity $Co_2(CO)_8$ is critical for successful catalytic reactions. A freshly opened sample of $Co_2(CO)_8$ stored in a dry box is often sufficiently pure to be used directly in this reaction provided three eq of cyclohexylamine per $Co_2(CO)_8$ are added to the reaction mixture. However, a more reliable approach is to generate the active cobalt catalyst in situ as described herein.

11. Ether was distilled from a purple solution of sodium/benzophenone ketyl and degassed by bubbling with argon.

12. Celite 545 was purchased from Fluka.

13. The bath temperature of the rotary evaporator must be kept below 35°C.

14. The checkers obtained 1.35 g (75%) Properties of 5[10]: ^{1}H NMR ($CDCl_3$) δ: 1.55 (s, 6H), 1.70 (s, 1 H), 6.01 (s, 1 H); ^{13}C NMR ($CDCl_3$) δ: 33.4, 71.7, 72.9, 76.8, 107.3, 200.0 (m).

15. Carbon monoxide is highly toxic and should be handled in a well-ventilated hood.

16. All solvents must be rigorously degassed for the Pauson-Khand step. 1,2-DME obtained from VWR Scientific was freshly distilled from a deep blue solution of sodium/benzophenone ketyl and degassed with argon prior to use.

17. The internal temperature of the mixture is 64°C. Temperatures above 70°C lead to catalyst decomposition.

18. To facilitate transfer of the catalyst, the CO source connected to the large reaction vessel is removed and replaced with a line leading to a mineral oil bubbler. When the transfer is complete, the bubbler is disconnected and a CO-filled double-walled balloon is connected to the top of the condenser and used to maintain a CO

atmosphere throughout the reaction. Strict exclusion of air during these transfers is critical.

19. The course of the reaction may be monitored by TLC: R_f of **3** and **6** is 0.82 and 0.36, respectively (1:1 EtOAc:hexanes). The color of the reaction mixture typically turns from burgundy to dark brown at the end of the reaction.

20. The solid residue is dissolved in a minimum of dichloromethane and purified by flash chromatography on silica gel (8 cm diameter by 15 cm long column), eluting with 1.5 % EtOH in CH_2Cl_2.

21. The crystals are collected by filtration on a fritted glass filter funnel, rinsed with ether and dried under reduced pressure.

22. The checkers obtained 4.3 g (65%). Properties of **6**: mp 147.0-148.0°C (2-PrOH/CH_2Cl_2); ^1H NMR ($CDCl_3$) δ: 1.96-2.03 (dd, J = 17.9, 3.7, 1H), 2.37 (s, 3H), 2.48-2.60 (m, 2H), 3.04-3.10 (m, 1H), 3.93-4.00 (dd, J = 14.3, 5.1, 2H), 4.27 (d, J = 16.5, 1H), 5.92 (s, 1H), 7.29 (d, J = 8.1, 2H), 7.66 (d, J = 8.2, 2H); ^{13}C NMR ($CDCl_3$) δ: 21.0 (CH_3), 39.3 (CH_2), 43.4 (CH), 47.1 (CH_2), 51.9 (CH_2), 125.6 (CH), 126.9(CH), 129.5 (CH), 133.0 (C), 143.6 (C), 178.2 (C), 206.8 (C); IR (KBr) cm^{-1}: 3067, 1711, 1648, 1343, 1162, 1090.

Waste Disposal Information

All toxic materials were disposed of in accordance with "Prudent Practices in the Laboratory"; National Academy Press; Washington, DC, 1995.

3. Discussion

The Pauson-Khand reaction is one of the most convenient methods for the synthesis of cyclopentenones, and both inter- and intramolecular variations exist.[3,4] In the classical procedure, the enyne is mixed with one equivalent of $Co_2(CO)_8$ and the mixture is heated (either at reflux or in a sealed tube) for several hours to days. The synthetic ease of the Pauson-Khand reaction was enhanced when it was discovered that silica gel, tertiary amine N-oxides and DMSO all markedly accelerated the reaction.[4] The Pauson-Khand reaction is compatible with enynes containing esters, ethers, amines, sulfides, 1,2-disubstituted olefins and substituted or terminal alkynes. A major drawback to large scale, classical Pauson-Khand reactions is the requirement for stoichiometric quantities of $Co_2(CO)_8$. We recently reported the first catalytic Pauson-Khand reaction conveniently operative at 1 atmosphere of CO.[5,6] The key factors of successful catalytic cyclizations are the use of highly purified $Co_2(CO)_8$, strictly anaerobic conditions, and temperature control. To circumvent the requirement for very pure $Co_2(CO)_8$ we developed a chemically robust, air-stable $Co_2(CO)_6$-alkyne complex as a source of pure $Co_2(CO)_8$ when released by silylative decomplexation.[7] This procedure employs cyclohexylamine, which has been shown to accelerate stoichiometric Pauson-Khand reactions[8] and minimizes the sensitivity of the catalyst to oxygen. The scope of the catalytic Pauson-Khand has been further enhanced by the discovery that 2,2,2-trifluoroethanol as a co-solvent minimizes some non-carbonylative isomerization processes.[9] The utility of the method is demonstrated by the diverse range of substrates shown in the Table.

TABLE
THERMALLY PROMOTED CATALYTIC PAUSON-KHAND CYCLIZATIONS

Entry	Enyne	Product	Yield (diastereomer ratio)
1	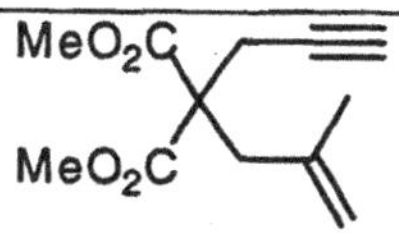 MeO_2C, MeO_2C	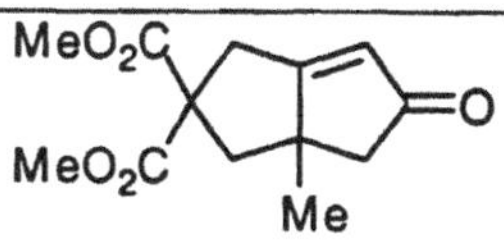 MeO_2C, MeO_2C, O, Me	95[a]
2	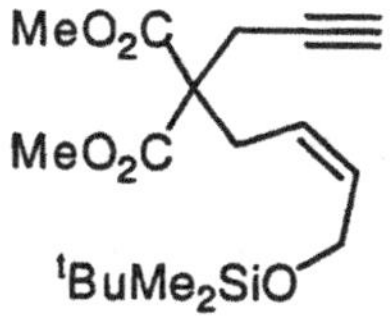 MeO_2C, MeO_2C, $^tBuMe_2SiO$	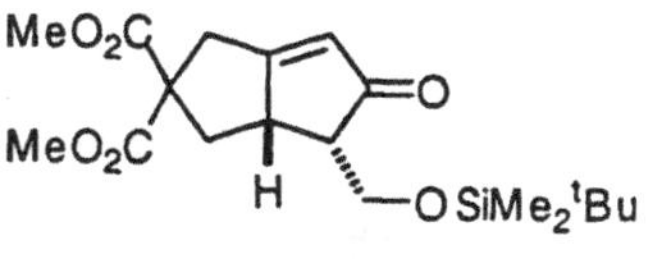 MeO_2C, MeO_2C, O, H, $OSiMe_2{}^tBu$	90[b] (>30:1)
3	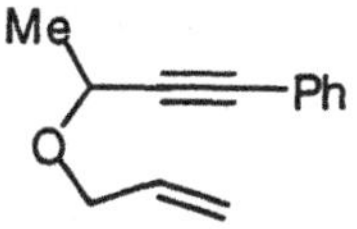 Me, Ph, O	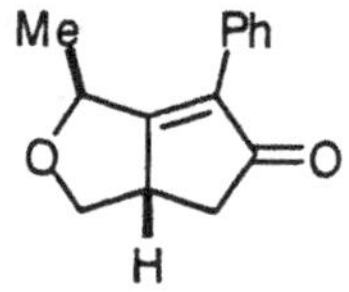 Me, Ph, O, O, H	96[a,c] (30:1)
4	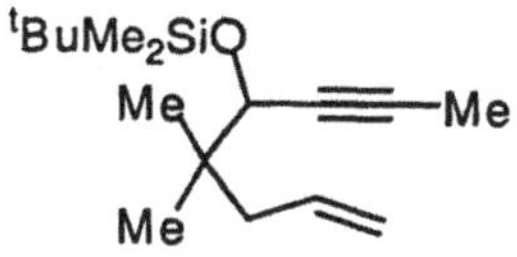 $^tBuMe_2SiO$, Me, Me, Me	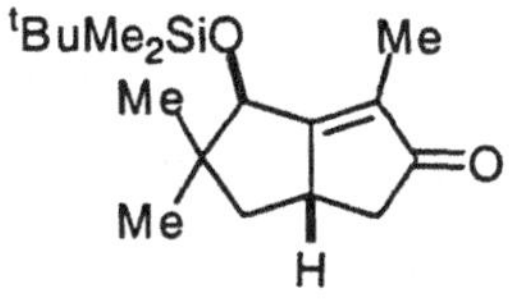 $^tBuMe_2SiO$, Me, Me, Me, O, H	74[a,d] (2.5:1)
5	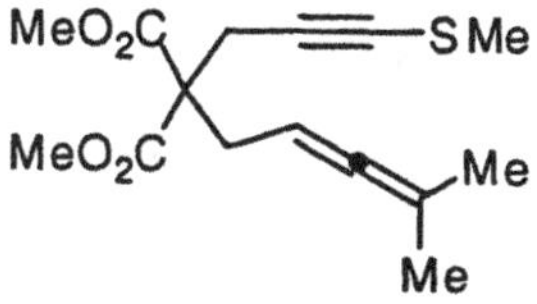 MeO_2C, MeO_2C, SMe, Me, Me	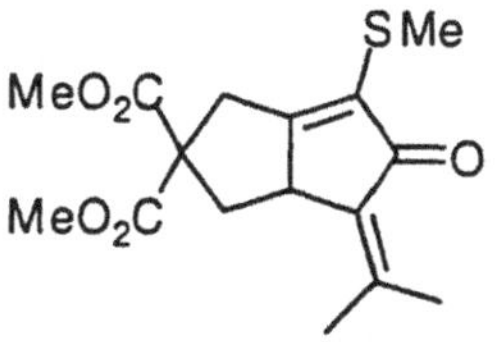 MeO_2C, MeO_2C, SMe, O	84[b]

[a]5 mol% $Co_2(CO)_8$. [b]10 mol% $Co_2(CO)_8$. [c]Reaction solvent was 2,2,2-trifluoroethanol; see ref. 9. [d]Reaction solvent was 2:1 2,2,2-trifluoroethanol:1,2-DME; see ref. 9.

1. The University of Texas at Austin, Department of Chemistry and Biochemistry, Austin, TX 78712.
2. Montana State University, Department of Chemistry and Biochemistry, Bozeman, MT 59717.
3. (a) Khand, I. U.; Knox, G. R.; Pauson, P.L.; Watts, W.E.; Foreman, M.I. *J. Chem. Soc., Perkin Trans. 1* **1973**, 977; (b) Pauson, P.L. *Tetrahedron* **1985**, *41*, 5855.
4. The Pauson-Khand reaction has been extensively reviewed: (a) Schore, N.E. *Chem. Rev.* **1988**, *88*, 1081; (b) Schore, N.E. *Org. React.* (NY) **1991**, *40*, 1; (c) Schore, N.E. In *Comprehensive Organic Synthesis;* Trost, B.M.; Fleming, I., Eds.; Pergamon: Oxford, 1991; Vol 5, p 1037; (d) Schore, N.E. In *Comprehensive Organometallic Chemistry II;* Hegedus, L.S., Ed. Elsevier: Oxford, 1995; Vol. 12, p. 703; (e) Brummond, K. M.; Kent, J. L. *Tetrahedron* **2000**, *56*, 3263.
5. Pagenkopf, B. L.; Livinghouse, T. *J. Am. Chem. Soc.* **1996**, *118*, 2285.
6. Belanger, D. B.; O'Mahony, D. J. R.; Livinghouse, T. *Tetrahedron Lett.* **1998**, *39*, 7637.
7. Belanger, D. B.; Livinghouse, T. *Tetrahedron Lett.* **1998**, *39*, 7641.
8. Sugihara, T.; Yamada, M.; Ban, H.; Yamaguchi, M.; Kaneko, C. *Angew. Chem. Int. Ed. Engl.* **1997**, *36*, 2801.
9. Pagenkopf, B. L.; Belanger, D. B.; O'Mahony, D. J. R.; Livinghouse, T. *Synthesis* **2000**,1009.
10. Piper, J. R.; Rose, L. M.; Johnston, T. P.; Grennan, M. M. *J. Med. Chem.* **1975**, *18*, 803.
11. Giordano, R.; Sappa, E.; Predieri, G. *Inorg. Chim. Acta* **1995**, *228*, 139.

Appendix

Chemical Abstracts Nomenclature (Collective Index Number); (Registry Number)

N-(2-Propenyl)-4-methylbenzenesulfonamide: Benzenesulfonamide, 4-methyl-N-2-propenyl- (9); (50487-71-3)

p-Toluenesulfonyl chloride: Benzenesulfonyl chloride, 4-methyl- (9); (98-59-9)

Allylamine: 2-Propen-1-amine (9); (107-11-9)

Pyridine (8, 9); (110-86-1)

N-(2-Propenyl)-N-(2-propynyl)-4-methylbenzenesulfonamide: Benzenesulfonamide, 4-methyl-N-2-propenyl-N-2-propynyl- (9); (133886-40-5)

1-Bromo-2-propyne: 1-Propyne, 3-bromo- (9); (106-96-7)

Hexacarbonyl[μ[(3,4-η:3,4-η)-2-methyl-3-butyn-2-ol]]dicobalt: Cobalt, hexacarbonyl [μ-[(3,4-η:3,4-η)-2-methyl-3-butyn-2-ol]]di-, (Co-Co) (9); (40754-33-4)

Dicobalt octacarbonyl: Cobalt, di-μ-carbonylhexacarbonyl di-, (Co-Co) (8, 9); (10210-68-1)

2-Methyl-3-butyn-2-ol: 3-Butyn-2-ol, 2-methyl- (8, 9); (115-19-5)

2,3,3α,4-Tetrahydro-2-[(4-methylbenzene)sulfonyl]cyclopenta[c]pyrrol-5(1H)-one: Cyclopenta[b]pyrrol-5(1H)-one, 2,3,3a,4-tetrahydro-1-[(4-methylphenyl)sulfonyl]- (9); (205885-50-3)

Carbon monoxide (8, 9); (630-08-0)

1,2-Dimethoxyethane: Ethane, 1,2-dimethoxy- (8, 9); (110-71-4)

Cyclohexylamine: Cyclohexanamine (9); (108-91-8)

Triethylsilane: Silane, triethyl- (8, 9); (617-86-7)

RUTHENIUM-CATALYZED ALKYLATION OF AROMATIC KETONES WITH OLEFINS: 8-[2-(TRIETHOXYSILYL)ETHYL]-1-TETRALONE

(1(2H)-Naphthalenone, 3,4-dihydro-8-[2-(triethoxysilyl)ethyl]-)

$$\text{1-tetralone} + \text{CH}_2\text{=CHSi(OEt)}_3 \xrightarrow[\text{toluene, reflux}]{\text{cat. RuH}_2\text{(CO)(PPh}_3\text{)}_3} \text{8-[2-(Si(OEt)}_3\text{)ethyl]-1-tetralone}$$

Submitted by Fumitoshi Kakiuchi and Shinji Murai.[1]

Checked by Dennis P. Curran and Andre Lapierre.

1. Procedure

8-[2-(Triethoxysilyl)ethyl]-1-tetralone (Note 1). An apparatus, consisting of a 100-mL, two-necked, round-bottomed flask, reflux condenser connected to a vacuum/N_2 line, inlet tube sealed with a rubber septum, and magnetic stirring bar, is evacuated, then flushed with nitrogen. This cycle is repeated four times. The apparatus is flame-dried under a flow of nitrogen, then cooled to room temperature under a nitrogen atmosphere. Carbonyldihydridotris(triphenylphosphine)ruthenium(II), $RuH_2(CO)(PPh_3)_3$, (0.918 g, 1.00 mmol, 0.010 eq) (Notes 2, 3) is placed in the flask under a slow flow of nitrogen. Addition of 20 mL of toluene (Note 4) to the flask affords a suspension of white solids (Note 5). To this suspension are added triethoxyvinylsilane (20.93 g, 110 mmol, 1.10 eq) (Notes 6, 7), then 1-tetralone (14.62 g, 100 mmol, 1.00 eq) (Notes 6, 8) at room temperature through the rubber septum using a syringe. The resulting mixture containing the white solids is heated to reflux in an oil bath (oil bath temperature: 135°C) (Notes 9, 10). The reaction mixture becomes colorless within 1 min, then changes to a dark wine red color within 5 min (Note 11). After heating for 30 min, the reaction mixture is cooled to room temperature. About half of the reaction mixture is transferred to a 50-mL, round-bottomed flask (Note 12) for concentration and distillation. Volatile materials (toluene and triethoxyvinylsilane) are removed by rotary evaporation under reduced pressure (2 mm) at 40°C. After almost all volatile materials are removed, the other half of the reaction mixture is transferred to the flask. The reaction vessel is rinsed with two 5-mL portions of toluene, then the rinses are transferred to the distillation flask. The combined reaction mixture and rinses are concentrated. Solvent and vinylsilane are removed by rotary evaporation under reduced pressure (2 mm at 40°C) (Note 13). Distillation of the residue under reduced pressure gives 31-32.5 g (92-96%) of 8-[2-(triethoxysilyl)ethyl]-1-tetralone as a pale yellow liquid, bp 133-135°C/0.2 mm (Notes 14-18).

2. Notes

1. This procedure is a modification of one published by our group.[2,3]

2. The ruthenium complex, carbonyldihydridotris(triphenylphosphine)ruthenium(II), can be synthesized according to the literature method.[3,4] This complex is commercially available from Aldrich Chemical Company, Inc. and Strem Chemicals, Inc.

3. The commercially available ruthenium complex (Strem Chemicals, Inc.) and the complex prepared in the submitters' laboratory showed comparable catalytic activity. The checkers purchased the complex from Strem Chemicals, Inc.

4. Toluene was dried over CaH_2, then distilled under a N_2 atmosphere.

5. The ruthenium complex is only slightly soluble in toluene at room temperature.

6. Triethoxyvinylsilane, 1-tetralone, and toluene were purchased from commercial suppliers (Aldrich Chemical Company, Inc., Strem Chemicals, Inc., Chisso Co., or Wako Pure Chemical Industries, Ltd.).

7. The silane was dried over CaH_2, then distilled under reduced pressure (bp = 70°C/6 mm).

8. 1-Tetralone was dried over $CaSO_4$, then distilled under reduced pressure (bp 85°C/2 mm).

9. Maintaining the oil bath temperature above 130°C (preferentially around 135°C) is necessary to effect the catalytic reaction with good reproducibility.

10. The actual temperature of the reaction mixture was around 125°C throughout the reaction.

11. The color of the reaction mixture changes to dark green, wine red, and finally dark wine red within 5 min.

12. The use of a small flask is recommended to minimize loss of the product.

13. The checkers transferred the entire reaction mixture to a 100-mL, round-bottomed flask with the aid of 2 x 5-mL washes with toluene. A distillation apparatus with a three-way adapter and three collection vessels was attached, and the solvent was removed by gentle heating at 3 mm. The residue was then distilled as described.

14. The second fraction is collected. The first fraction (0.20-0.28 g; 65-100°C/0.2 mm) containing unknown impurities and a small amount of the desired product is discarded.

15. The residue is prone to bump during the late stages of the distillation. Wine red splashes on the wall of the distillation head results in contamination of the product with impurities, including triphenylphosphine, triphenylphosphine oxide, and the ruthenium complex. If the product is contaminated with these impurities, it can be redistilled.

16. Heating the distillation head with a heat gun is recommended to avoid reducing the yield of the product.

17. The spectral properties are as follows: ^1H NMR (300 MHz, $CDCl_3$) δ: 0.94-1.00 (m, 2 H, $SiCH_2$), 1.26 (t, 9 H, J = 6.97, CH_3), 2.05 (quintet, 2 H, J = 6.00, $CH_2\underline{CH_2}CH_2$), 2.61 (t, 2 H, J = 6.33, $ArCH_2$), 2.90 (t, 2 H, J = 5.66, $C(O)CH_2$), 3.08-3.14 (m, 2 H, $Ar\underline{CH_2}CH_2Si$), 3.87 (q, 6 H, J = 5.58, OCH_2), 7.05 (d, 1 H, J = 7.17, ArH), 7.11 (d, 1 H,

J = 7.47, ArH), 7.26-7.31 (dd, 1 H, J = 7.56, ArH); ^{13}C NMR (75 Hz, $CDCl_3$) δ: 12.18 ($SiCH_2$), 18.21(CH_3), 22.81 (CH_2), 28.57, 30.82 [($SiCCH_2$) and ($ArCH_2$)], 40.96 ($C(O)CH_2$), 58.18 (OCH_2), 126.73, 129.15, 130.31, 132.33, 145.68, 147.95 (Ar), 199.60 (C=O); MS (m/z): 336 (M^+), 290 (M^+ - EtOH), 189, 173, 163, 135, 115, 79, 63; IR (neat) cm^{-1}: 1680 (C=O). Anal. Calcd for $C_{18}H_{28}O_4Si$: C, 64.25; H, 8.39. Found: C, 64.24; H, 8.29.

18. The checkers obtained bp 160-168°C/2 mm.

Waste Disposal Information

All toxic materials were disposed of in accordance with "Prudent Practices in the Laboratory"; National Academy Press; Washington, DC, 1995.

3. Discussion

The procedure described here is typical for the catalytic alkylation of aromatic ketones at the ortho position by alkenes. Aromatic ketones are readily available by Friedel-Crafts acylation and many other methods,[5] and many of these ketones are suitable substrates for the present catalytic alkylation with alkenes affording the corresponding ortho-alkylated ketones.[3,6] The present method provides a direct way to alkylate aromatics with olefins. Moreover, the C-C bond formation takes place with exclusive ortho selectivity, while mixtures of o-, m-, p-isomers are usually obtained in the conventional Friedel-Crafts alkylation of aromatic compounds.

In the present ruthenium-catalyzed reaction, a C-C bond is formed directly from a C-H bond without prior conversion of the C-H bond to another functional group such as halogen. The preparation of 8-[2-(triethoxysilyl)ethyl]-1-tetralone described above can be modified for a variety of substituted aromatic ketones. Some additional, representative examples of the $RuH_2(CO)(PPh_3)_3$-catalyzed addition of C-H bonds in aromatic ketones to olefins are shown in the Table.[3,6] Naphthalene derivatives and heteroaromatic ketones can be employed in the present coupling reaction. Functional group compatibility of this reaction is broad and both electron-donating and electron-withdrawing substituents are tolerated. In many cases, simple bulb-to-bulb distillation of the reaction mixture gives an analytically pure product. This simple one-pot procedure provides a new opportunity for site-selective alkylations of aromatic and heteroaromatic compounds.

TABLE RUTHENIUM-CATALYZED REACTION OF AROMATIC KETONES WITH OLEFINS

Entry	Ketone	Product	Time	Isolated Yield	GLC Yield
1[a,b]	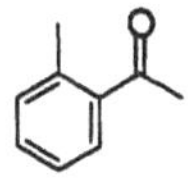	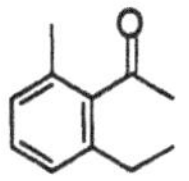	24 h	88%	100%
2[a,c]	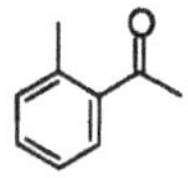	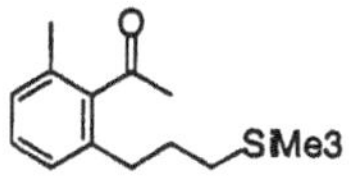 SiMe3	4 h	97%	100%
3[a,d]	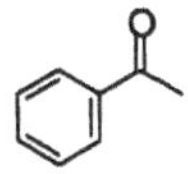	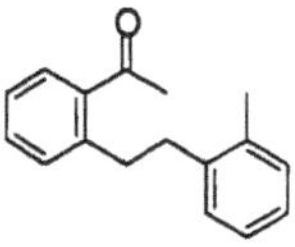	4 h	89%	—
4[a,e]	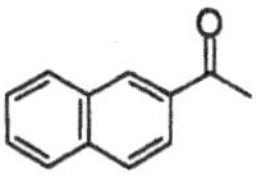	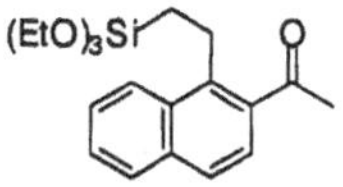 (EtO)$_3$Si	6 h	88%	100%
5[a,e]	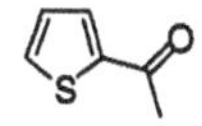 S	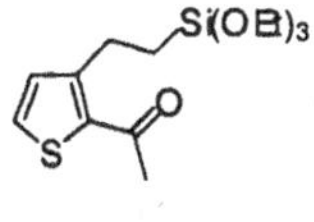 Si(OEt)$_3$ S	1 h	93%	100%
6[e,f]	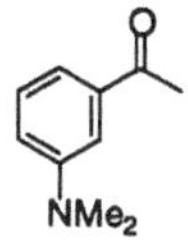 NMe$_2$	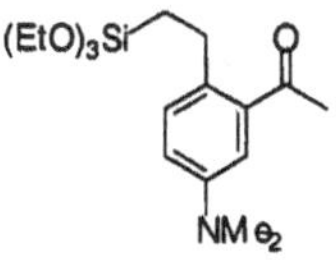 (EtO)$_3$Si NMe$_2$	8 h	72%	85%
7[e,f]	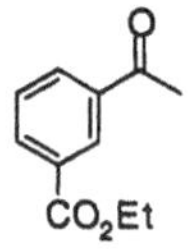 CO$_2$Et	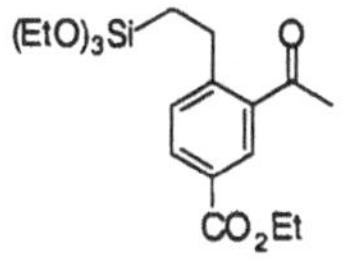 (EtO)$_3$Si CO$_2$Et	4 h	91%	91%
8[e,f]	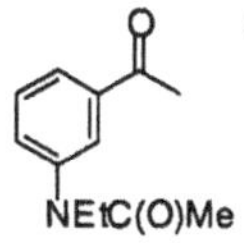 NEtC(O)Me	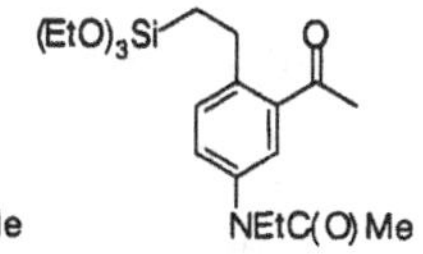 (EtO)$_3$Si NEtC(O)Me	8 h	92%	96%
9[e,f]	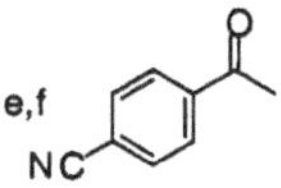 NC	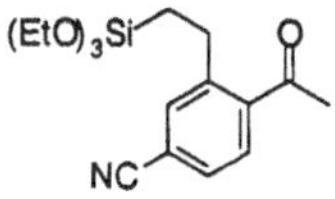 (EtO)$_3$Si NC	5 h	73%	88%

[a]See ref 3. [b]Ethylene was used. [c]Allyltrimethylsilane was used. [d]2-Methylstyrene was used. [e]Triethoxyvinylh

1. Department of Applied Chemistry, Faculty of Engineering, Osaka University, Suita, Osaka 565-0871, Japan.
2. Murai, S.; Kakiuchi, F.; Sekine, S.; Tanaka, Y.; Kamatani, A.; Sonoda, M.; Chatani, N. *Nature* **1993**, *366*, 529.
3. Kakiuchi, F.; Sekine, S.; Tanaka, Y.; Kamatani, A.; Sonoda, M.; Chatani, N.; Murai, S. *Bull. Chem. Soc. Jpn.* **1995**, *68*, 62.
4. Ahmad, N.; Levison, J. J.; Robinson, S. D.; Uttley, M. F. *Inorg. Synth.* **1974**, *15*, 45.
5. Walter, D. S. In "Comprehensive Organic Functional Group Transformations"; Katritzky, A. R.; Meth-Cohn, O.; Rees, C. W., Eds.; Pergamon: Cambridge, U.K., 1995; Vol. 3, Chapter 3.06.
6. Sonoda, M.; Kakiuchi, F.; Chatani, N.; Murai, S. *Bull. Chem. Soc. Jpn.* **1997**, *70*, 3117.

Appendix

Chemical Abstracts Nomenclature (Collective Index Number); (Registry Number)

8-[2-(Triethoxysilyl)ethyl-1-tetralone: 1(2H)-Naphthalenone, 3,4-dihydro-8-[2-(triethoxysilyl)ethyl]- (9); (154735-94-1)

Carbonyldihydridotris(triphenylphosphine)ruthenium(II): Ruthenium, carbonyldihydridotris(triphenylphosphine) (8,9); (25360-32-1)

Triethoxyvinylsilane: Silane, ethenyltriethoxy- (9); (78-08-0)

1-Tetralone: 1(2H)-Naphthalenone, 3,4-dihydro- (8,9); (529-34-0)

INTRA- AND INTERMOLECULAR KULINKOVICH CYCLOPROPANATION REACTIONS OF CARBOXYLIC ESTERS WITH OLEFINS: BICYCLO[3.1.0]HEXAN-1-OL AND trans-2-BENZYL-1-METHYLCYCLOPROPAN-1-OL

(Bicyclo[3.1.0]hexan-1-ol and 1-Cyclopropanol, 1-methyl-2-phenylmethyl-)

A. Methyl 5-hexenoate (CO_2Me) → (n-BuMgCl, $(i\text{-}PrO)_3TiCl$, ether) → **1**

B. Allylbenzene (Ph) + CH_3CO_2Et → ($c\text{-}C_5H_9MgCl$, $(i\text{-}PrO)_3TiCl$, THF) → **2**

Submitted by Se-Ho Kim, Moo Je Sung, and Jin Kun Cha.[1]

Checked by Fanglong Yang and Dennis P. Curran.

1. Procedure

A. *Bicyclo[3.1.0]hexan-1-ol* (**1**). To a 500-mL, round-bottomed flask, equipped with a magnetic stirring bar and rubber septum (Note 1), is added at room temperature a mixture of 2.0 g (15.6 mmol) of methyl 5-hexenoate, 11.2 mL (11.2 mmol) of a 1 M solution of chlorotitanium triisopropoxide in hexane, and 54 mL of anhydrous ether (Note 2) under a nitrogen atmosphere. A 1M solution of n-butylmagnesium chloride in ether (52 mL, 52 mmol; Note 3) is added over a period of 6.5 hr via a syringe pump (Note 4) at room temperature. After the addition is complete, the resulting black reaction mixture is stirred for an additional 20 min. The mixture is cooled to 0°C with an ice bath, diluted with 50 mL of ether and then quenched by slow addition of water (14 mL). The resulting mixture is stirred for an additional 3 hr at room temperature. The organic phase is separated and the aqueous phase is extracted with ether (3 x 100 mL). The combined organic extracts are washed with brine (2 x 50 mL), dried over anhydrous magnesium sulfate, filtered, and concentrated under reduced pressure using a rotary evaporator (Note 5). Purification of the crude product (Note 6) by column chromatography on 40 g of silica gel (Note 7) using a gradient of 5% to 10% ether/pentane as the eluent provides 1.09 g (71%) of bicyclo[3.1.0]hexan-1-ol (**1**)[2] as a colorless oil (Note 8).

B. *trans-2-Benzyl-1-methylcyclopropan-1-ol* (**2**). To a 500-mL, round-bottomed flask, equipped with a magnetic stirring bar and rubber septum, is added a mixture of 2.5 g (21 mmol) of allylbenzene, 2 mL (20 mmol) of ethyl acetate, and 20 mL of a 1M solution of chlorotitanium triisopropoxide in hexane (Note 2), and 160 mL of anhydrous tetrahydrofuran (THF). After the mixture has been cooled to 0°C with an ice bath under a nitrogen atmosphere, a 1M solution of cyclopentylmagnesium chloride in ether (80 mL, 80 mmol; Note 9) is added over a period of 2.5 hr via a syringe pump (Note 4). After the addition is complete, the resulting black reaction mixture is stirred for 30 min at 0°C, then is quenched by the cautious addition of water (15 mL). The resulting mixture is stirred for an additional 1 hr at room temperature and filtered through a pad of Celite, which is rinsed thoroughly with ether (4 x 50 mL). The combined filtrate and rinsings are poured into a separatory funnel containing 50 mL of water and shaken thoroughly. The organic phase is separated, washed with brine (50 mL), dried over anhydrous magnesium sulfate, filtered, and concentrated under reduced pressure using a rotary evaporator. Purification of the crude product (obtained as a pale yellow oil) by column chromatography on 80 g of silica gel (Note 7) using 1:20 ethyl acetate:hexane as the eluent provides 2.6 g (80%) of trans-2-benzyl-1-methylcyclopropan-1-ol (**2**) as a colorless oil (Note 10).

2. Notes

1. All glassware, needles, syringes were oven-dried (>100°C) and quickly assembled prior to use.

2. (a) Methyl 5-hexenoate was purchased from TCI America and used as received. A 1M solution of chlorotitanium triisopropoxide in hexane was purchased from Aldrich Chemical Company, Inc. and used as obtained. (b) Ether and THF were dried and purified by distillation from sodium/benzophenone. Unless stated otherwise, all solvents and reagents in this procedure were obtained commercially and used as received.

3. n-Butylmagnesium chloride (a 2M solution in ether) was available from Aldrich Chemical Company, Inc. Alternatively, it was freshly prepared from 1-chlorobutane and magnesium.

4. (a) The tip of a syringe needle was placed right above the surface of the reaction mixture, rather than near the neck of the reaction flask. This placement was found to minimize the amount of the Grignard reagent to be added. An excess of the Grignard reagent was used to ensure complete conversion. (b) To prevent the glass syringe plunger from sticking in the barrel, vacuum grease was applied around the edge of the exposed portion of the plunger.

5. The solvent was removed under water (aspirator) vacuum, while the bath temperature was kept below 20°C. Otherwise, considerable loss occurred due to the volatility of the cyclopropanol product.

6. Immediate purification of the crude product is recommended. Otherwise, it should be stored in a cold (–30°C) freezer to avoid decomposition.

7. Flash chromatography was performed on E. Merck silica gel (70-230 mesh). For Procedure A, the checkers used 60 g silica gel, eluting first with 150 mL 10% ether/pentane, then with 200 mL 20% ether/pentane. For Procedure B, the checkers used 100 g silica gel and eluted first with 100 mL 9% ether/hexane, then with 200 mL 17% ether/hexane.

8. Yields in several runs were obtained in the range of 68–75% by the submitters and 67-71% by the checkers. Spectral data for **1** [R_f = 0.27 (1:4 ethyl acetate:hexane)] are as follows: bp 53-55°C (15 mm, Kugelrohr) [lit.[2b] bp 63-65°C (20 mm)]; IR (neat) cm^{-1}: 3300; ^1H NMR (360 MHz, acetone-d_6) δ 0.47 (t, J = 4.6, 1 H), 0.67 (dd, J = 4.6, 8.9, 1 H), 1.02–1.15 (m, 2 H), 1.43 (dd, J = 7.9, 12.1, 1 H), 1.55 (m, 1 H), 1.75–1.89 (m, 3 H), 4.25 (s, 1 H, -OH); ^{13}C NMR (90 MHz, acetone-d_6) δ 14.9, 21.8, 24.3, 27.4, 34.3, 64.2; Mass spectrum: m/z 98 (M^+, 29), 97 (47), 83 (56), 70 (100), 55 (54).

9. Cyclopentylmagnesium chloride (a 2M solution in ether) was available from Aldrich Chemical Company, Inc. Alternatively, it was freshly prepared from chlorocyclopentane and magnesium.

10. Yields in several runs were obtained in the range of 78–86% by the submitters and 78-82% by the checkers. Spectral data for **2** [R_f = 0.47 (1:3 ethyl acetate:hexane)] are as follows: IR (neat) cm^{-1}: 3332, 1228; ^1H NMR (360 MHz, $CDCl_3$) δ 0.29 (dd, J = 5.4, 6.0, 1 H), 0.98 (dd, J = 5.4, 10.0, 1 H), 1.30–1.39 (m, 1 H), 1.49 (s, 3 H), 1.88 (s, 1 H, -OH), 2.55 (dd, J = 7.5, 15.3, 1 H), 2.67 (dd, J = 7.3, 15.3, 1 H), 7.19–7.33 (m, 5 H); ^{13}C NMR (90 MHz, $CDCl_3$) δ 20.4, 20.8, 26.1, 35.6, 55.8, 125.9, 128.1, 128.4, 141.6; Mass spectrum (EI): m/z 162 (M^+, 3), 104 (76), 92 (58), 71 (100).

Waste Disposal Information

All toxic materials were disposed of in accordance with "Prudent Practices in the Laboratory"; National Academy Press: Washington, DC, 1995.

3. Discussion

Kulinkovich and co-workers recently developed a new method for preparing cyclopropanols (the subject of recent reviews[3]) by treating carboxylic esters with a suitable Grignard reagent (i.e., containing β-hydrogen atoms) in the presence of titanium tetraisopropoxide.[4] The present procedures represent a convenient variant of the Kulinkovich cyclopropanation reaction by using a sacrificial Grignard reagent (n-butylmagnesium chloride or cyclopentylmagnesium chloride) to generate the putative titanacyclopropane intermediate for subsequent ligand exchange with a monosubstituted olefin. A mixture of the two rapidly interconverting titanacyclopropane intermediates is thus formed, followed by the intramolecular cyclopropanation reaction, to afford the corresponding bicyclic cyclopropanol products in good to excellent yield.[5,6] Because formation of five- and six-membered rings proceeds much faster than the intermolecular reaction, the precise position of equilibrium is inconsequential for successful cyclization, as demonstrated by the facile preparation of bicyclo[3.1.0]hexan-1-ol (**1**). Consequently, various Grignard reagents have been shown to be effective for the intramolecular cyclopropanation reactions with insignificant differences in yields. Whereas several titanium alkoxides and aryloxides can also be employed, chlorotitanium triisopropoxide and methyltitanium triisopropoxide have often been found to be the titanium reagent of choice.[7] Ether, THF, toluene, or even dichloromethane are generally appropriate reaction solvents.

$Ti(OPr\text{-}i)_2$ ← $ClTi(OPr\text{-}i)_3$ + n-BuMgCl (2 equiv)

β-elimination

[$Ti(OPr\text{-}i)_2$] + CO_2Me -- slower -->

⇅

+ $(i\text{-}PrO)_2Ti$ CO_2Me -- faster --> OH H 1

Central to the successful extension of the olefin exchange strategy to intermolecular coupling reactions of an ester and a monosubstituted olefin, as illustrated by the preparation of trans-2-benzyl-1-methylcyclopropan-1-ol (2), is the use of cyclohexyl or cyclopentyl Grignard reagents.[8] These secondary Grignard reagents are believed to allow the position of equilibrium to be shifted to the desired titanacyclopropane intermediate (less substituted, more stable). Notable characteristics of the ligand exchange-mediated variant of the Kulinkovich cyclopropanation include the simplicity and ease of operation using inexpensive reagents and also excellent compatibility with several common functional groups (for example, alkoxy, bromo, and alkenyl substituents).

In addition, other carboxylic acid derivatives, such as tertiary carboxamides and cyclic carbonates, readily undergo cyclopropanation reactions under similar conditions to provide cyclopropylamines[9,10] and cyclopropanone hemiacetals,[11] respectively. The cognate titanium-mediated coupling of imides has been shown to afford synthetically useful N-acylhemiaminals.[12]

1. Department of Chemistry, University of Alabama, Tuscaloosa, AL 35487.
2. Bicyclo[n.1.0]alkan-1-ols were previously prepared by the Simmons-Smith cyclopropanation reaction of 1-(trimethylsilyloxy)cycloalkenes and subsequent hydrolysis: (a) Conia, J. M.; Girard, C. *Tetrahedron Lett.* **1973**, 2767. (b) Murai, S.; Aya, T.; Sonoda, N. *J. Org. Chem.* **1973**, *38*, 4354.
3. (a) Kulinkovich, O. G.; Sviridov, S. V.; Vasilevskii, D. A.; Pritytskaya, T. S. *Zh. Org. Khim.* **1989**, *25*, 2244. (b) Kulinkovich, O. G.; Sviridov, S. V.; Vasilevskii, D. A.; Savchenko, A. I.; Pritytskaya, T. S. *Zh. Org. Khim.* **1991**, *27*, 294. (c) Kulinkovich, O. G.; Sorokin, V. L.; Kel'in, A. V. *Zh. Org. Khim.* **1993**, *29*, 66. (d) Kulinkovich, O. G.; Savchenko, A. I.; Sviridov, S. V.; Vasilevskii, D. A. *Mendeleev Commun.* **1993**, 230.
4. (a) Kulinkovich, O. G.; de Meijere, A. *Chem. Rev.* **2000**, *100*, 2789. (b) Sato, F.; Urabe, H.; Okamoto, S. *Chem. Rev.* **2000**, *100*, 2835. (c) For a recent theoretical study on mechanism, see: Wu, Y.-D.; Yu, Z.-X. *J. Am. Chem Soc.* **2001**, *123*, 5777.
5. Lee, J.; Kang, C. H.; Kim, H.; Cha, J. K. *J. Am. Chem. Soc.* **1996**, *118*, 291.
6. (a) Kasatkin, A.; Sato, F. *Tetrahedron Lett.* **1995**, *36*, 6079. (b) Kasatkin, A.; Kobayashi, K.; Okamoto, S.; Sato, F. *Tetrahedron Lett.* **1996**, *37*, 1849.
7. (a) Corey, E. J.; Rao, S. A.; Noe, M. C. *J. Am. Chem. Soc.* **1994**, *116*, 9345. (b) Chaplinski, V.; Winsel, H.; Kordes, M.; de Meijere, A. *Synlett* **1997**, 111. (c) Lee, J. C.; Sung, M. J.; Cha, J. K. *Tetrahedron Lett.* **2001**, *42*, 2059.
8. Lee, J.; Kim, H.; Cha, J. K. *J. Am. Chem. Soc.* **1996**, *118*, 4198.
9. Chaplinski, V.; de Meijere, A. *Angew. Chem., Int. Ed. Engl.* **1996**, *35*, 413.
10. Lee, J.; Cha, J. K. *J. Org. Chem.* **1997**, *62*, 1584.
11. Lee, J.; Kim, Y. G.; Bae, J. G.; Cha, J. K. *J. Org. Chem.* **1996**, *61*, 4878.
12. Lee, J.; Ha, J. D.; Cha, J. K. *J. Am. Chem. Soc.* **1997**, *119*, 8127.

Appendix

Chemical Abstracts Nomenclature (Collective Index Number); (Registry Number)

Bicyclo[3.1.0]hexan-1-ol (8, 9); (7422-09-5)

Methyl 5-hexenoate: 5-Hexenoic acid, methyl ester (8, 9); (2396-80-7)

Chlorotitanium triisopropoxide: Titanium, chlorotris(2-propanolato)-, (T-4)- (9); (20717-86-6)

n-Butylmagnesium chloride: Magnesium, butylchloro- (8, 9); (693-04-9)

Allylbenzene: Benzene, 2-propenyl- (9); (300-57-2)

Cyclopentylmagnesium chloride: Magnesium, chlorocyclopentyl- (8, 9); (32916-51-1)

GENERATION OF AN ACETYLENE-TITANIUM ALKOXIDE COMPLEX: PREPARATION OF (Z)-1,2-DIDEUTERIO-1-(TRIMETHYLSILYL)-1-HEXENE

(Silane, (1,2-dideuterio-1-hexenyl)trimethyl-, (Z)-)

$$SiMe_3{-}C{\equiv}C{-}C_4H_9 \; + \; Ti(OPr\text{-}i)_4 \xrightarrow{i\text{-}PrMgCl} \left[Me_3Si,\ C_4H_9 \triangleright Ti(OPr\text{-}i)_2 \right] \xrightarrow{D_2O} Me_3Si(D)C{=}C(D)C_4H_9$$

Submitted by Hirokazu Urabe,[1a] Daisuke Suzuki,[1b] and Fumie Sato.[1b]

Checked by Katherine Hervert and Louis Hegedus.

1. Procedure

An oven-dried, 1-L, two-necked, round-bottomed flask is equipped with a stirring bar and 100-mL, pressure-equalizing addition funnel, the top of which is fitted with a rubber septum. The other neck of the flask is fitted with a three-way stopcock, one outlet of which is capped with a rubber septum and the other is connected alternately to a vacuum/argon gas system. The flask is evacuated, then flushed with argon; this operation is repeated twice. After 400 mL of anhydrous diethyl ether (Note 1), 1-(trimethylsilyl)-1-hexyne (Note 2) (5.00 g, 32.40 mmol) and freshly distilled titanium tetraisopropoxide (Note 3) (11.95 mL, 40.50 mmol) are added in this order through the septum of the three-way stopcock via syringe. The resulting solution is stirred and cooled in a dry ice-hexane bath to ca. –78°C. The addition funnel is charged via syringe through the rubber septum with a solution of isopropylmagnesium chloride in diethyl ether (Note 4) (50.62 mL of a 2.0M solution in ether, 101.24 mmol), which then is added dropwise to the cold mixture over 30 min. The mixture is warmed to –50°C over 30 min (and turns black) and is aged for 3 hr at –50°C to insure completion of reaction (Note 5).

ꞌeuterium oxide (12 mL) (Note 6) is rapidly added via the addition funnel to the cooled reaction mixture with stirring. The cooling bath is then removed and the reaction mixture is allowed to warm to room temperature. At this point, the addition funnel and three-way stopcock are removed. After anhydrous sodium sulfate (100 g) is added to the heterogeneous mixture, the suspension is stirred at room temperature for 10 min and filtered under vacuum through a short pad of Celite, which is washed with small portions of ether. The combined filtrate and washings are concentrated with a rotary evaporator (Note 7) to approximately 30 mL, which is transferred to a 50-mL distillation flask with the aid of small portions of ether. The crude product is purified by distillation under partial vacuum to furnish 3.56-4.47 g (69-87%) of the olefin, bp 128-129°C (350 mm), as a colorless liquid (Notes 8, 9).

2. Notes

1. Anhydrous diethyl ether was purchased from Kanto Chemical Co., Ltd. (Japan).

2. 1-(Trimethylsilyl)-1-hexyne was purchased from Aldrich Chemical Co., Inc. and used as received. Alternatively, this can be prepared according to a method described in Organic Syntheses.[2]

3. Titanium isopropoxide was purchased from Tokyo Kasei Kogyo Co., Ltd. (Japan), distilled before use, and stored under an argon atmosphere.

4. A 2.0M solution of isopropylmagnesium chloride in diethyl ether was purchased from Aldrich Chemical Co., Inc. and used as received.

5. Reduction of the amount of the ethyl ether (initially 400 mL) increases the formation of the 2:1 acetylene-titanium alkoxide complex, i.e., a titanacyclopentadiene. However, under these conditions, hydrolysis of the reaction mixture reveals that the formation of 2,3-dibutyl-1,4-bis(trimethylsilyl)-1,3-butadiene arising from the

titanacyclopentadiene is less than 3%. In any event, after distillation, this diene does not contaminate the desired product.

6. Deuterium oxide (100.0 atom% D) was purchased from Aldrich Chemical Co., Inc.

7. The product is moderately volatile so that extensive concentration under reduced pressure decreases the yield. When most of the solvent is removed under atmospheric pressure with a standard distillation apparatus (rather than with a rotary evaporator), the product is obtained in 87% yield.

8. The spectra are as follows: IR (neat) cm^{-1}: 2957, 2927, 2862, 1585 (C=C), 1459, 1248 (Si-Me), 839, 756. ^{1}H NMR (300 MHz, $CDCl_3$): δ 0.14 (s, 9 H, Me_3Si), 0.91 (t, J = 7.2, 3 H, Me), 1.35-1.55 (m, 4 H, $(CH_2)_2$), 2.22 (t, J = 7.2, 2 H, allylic CH_2). ^{13}C NMR (75 MHz, $CDCl_3$): δ 0.08 ($SiMe_3$), 13.91, 22,34, 31.90, 33.04, 128.23 (t, J ^{13}C-^{2}H = 20.5), 148.96 (t, J ^{13}C-^{2}H = 22.8). Anal. Calcd for $C_9H_{18}D_2Si$: C, 68.26. Found: C, 68.11. Deuterium incorporation α and β to the silyl group is determined by ^{1}H NMR analysis to be >99.5%.

9. Structural identification, except for the deuterium incorporation, is secured through (Z)-1-(trimethylsilyl)-1-hexene, a known compound obtained by hydrolysis of the acetylene-titanium complex with H_2O instead of D_2O using the same procedure. Spectra for the unlabeled product follow: IR (neat) cm^{-1}: 2958, 2925, 2859, 1607 (C=C), 1458, 1249 (Si-Me), 837, 763; ^{1}H NMR (300 MHz, $CDCl_3$): δ 0.11 (s, 9 H, Me_3Si), 0.90 (t, J = 7.2, 3 H, Me), 1.25-1.45 (m, 4 H, $(CH_2)_2$), 2.12 (symmetrical m, 2 H, allylic CH_2), 5.47 (d/t, J = 13.8, 1.2, 1 H, Me_3SiCH=C), 6.30 (d/t, J = 13.8, 7.2, 1 H, CH=$CHSiMe_3$). ^{13}C NMR (75 MHz, $CDCl_3$): δ 0.09 ($SiMe_3$), 13.90, 22.31, 31.99, 33.18, 128.80, 149.41. These data are in good agreement with reported values.[3] The ratio of Z and E isomers is determined to be >99.5:0.5 by ^{1}H NMR analysis of this sample. No contamination by the saturated silane is also confirmed by NMR analysis.

Waste Disposal Information

All toxic materials were disposed of in accordance with "Prudent Practices in the Laboratory"; National Academy Press; Washington, DC 1995.

3. Discussion

Isomerically pure 1-alkenylsilanes are useful intermediates in organic synthesis.[4] Thus, their deuterium-labeled counterparts are versatile precursors for the preparation of labeled organic molecules. Catalytic reduction of acetylenes using deuterium gas (D_2) is a convenient method to prepare D-labeled olefins.[5] However, some drawbacks have been observed, including scrambling between hydrogen atoms in the substrate and D_2 (which decreases the degree of deuterium uptake as well as the positional selectivity[6]), the variable E/Z ratio of the resulting olefinic bond,[7] and over-reduction of the acetylenic bond.[5] In addition, handling (and indeed availability) of the required amount of D_2 may be often problematic in some laboratories. The method described herein[8] is a convenient alternative to reduction with D_2 because (1) high deuterium incorporation of the vinylic positions and (2) formation of the Z-olefin is insured, and finally (3) inexpensive reagents (including D_2O as the source of deuterium) are utilized. The labeled olefins shown in the Table have been prepared by this method.[8,9]

Table. Preparation of (Z)-Dideuterio Alkenes

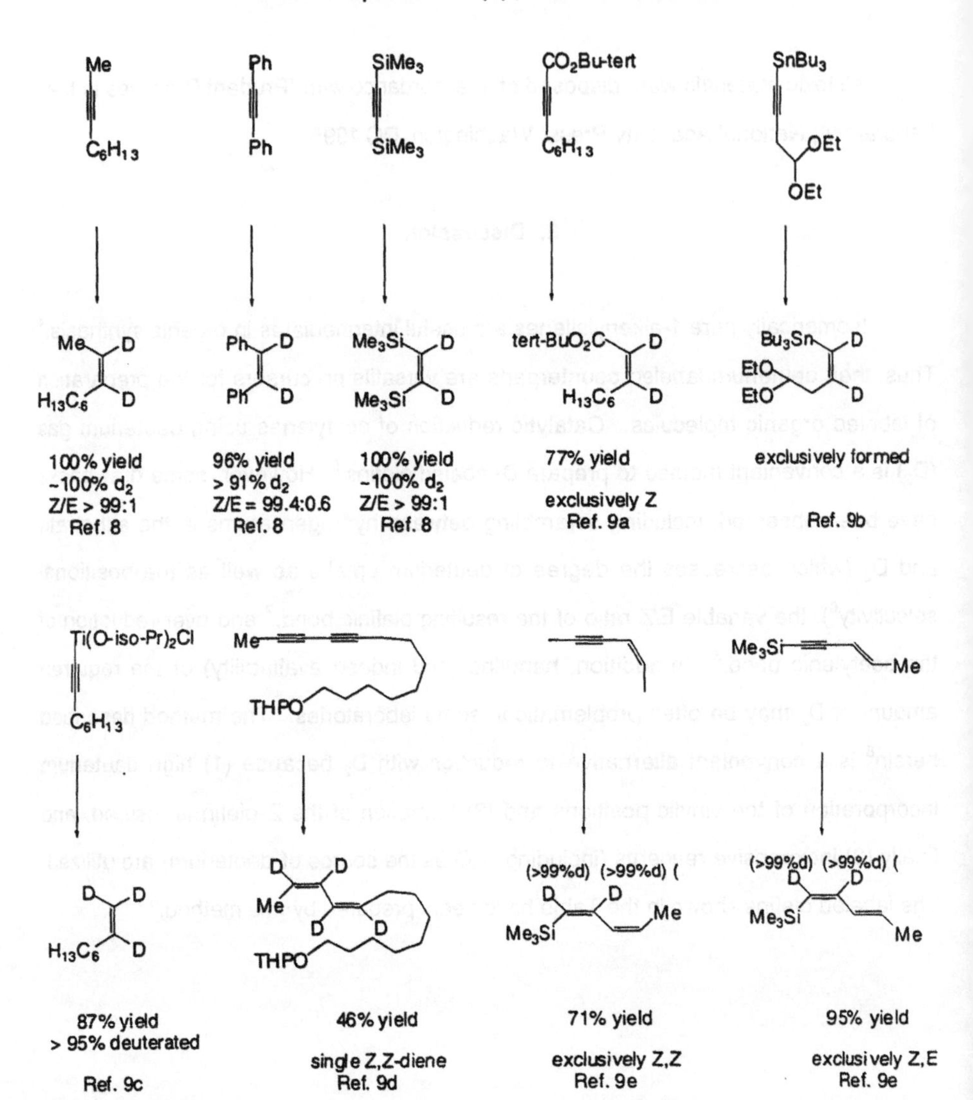

More importantly, the procedure represents a novel method for the generation of a new type of acetylene-Group 4 transition metal complex (**1** in the Scheme[10]). Acetylene-metal complexes often have been utilized as useful intermediates in organic synthesis, even though their structures may not be known.[11] The Scheme summarizes the four types of transformations available to the titanium complex **1**[10]: i) **1** can serve as a cis-1,2-bis-carbanionic species, which are otherwise difficult to generate, and affords adducts upon reaction with a variety of electrophiles (route a); ii) allenyl- or propargyltitanium reagents are generated from acetylenes having an appropriate leaving group at the propargylic position (route b); iii) the carbon-titanium bond of the complex reacts intramolecularly with an ester carbonyl group to give unsaturated carbonyl compounds (route c); iv) complex **1** participates in coupling reactions with other olefins or acetylenes to give the corresponding titanacycles, enabling a variety of subsequent synthetic applications (route d).

The synthetic advantages of the titanium alkoxide complex **1** over other Group 4 metal complexes such as titanocene or zirconocene-based ones are: i) the titanium alkoxide is very inexpensive compared with metallocene reagents; ii) the reaction work up is very simple, since titanium is completely hydrolyzed to give water-soluble inorganic salts, readily separable from the desired organic product(s); iii) various functionalized acetylene complexes have been generated for the first time; and finally, iv) several reactions, some of which cannot be accomplished by conventional Group 4 metal complexes, are now feasible through these new acetylene complexes.[10]

Scheme. Generation of and Synthetically Useful Reactions of Acetylene-Titanium Complexes 1

$R^1C{\equiv}CR^2$ + $Ti(OPr\text{-}i)_4$ + i-PrMgCl

R^1, R^2 = alkyl, aryl, silyl, heteroatom group, CO_2R, $CONR_2$

a: E^+-X^-

b: R^2 = CHXR

c: $R_2 = (CH_2)_3CO_2Et$

d: R–C≡C (alkyne)

1. (a) Department of Biological Information and (b) Department of Biomolecular Engineering, Graduate School of Bioscience and Biotechnology, Tokyo Institute of Technology, 4259 Nagatsuta-cho, Midori-ku, Yokohama-shi, Kanagawa 226-8501 Japan. Correspondence should be addressed to F. S.

2. Stang, P. J.; Kitamura, T. *Org. Synth., Coll. Vol. IX*, 1998, 477.

3. (a) Soderquist, J. A. ; Santiago, B. *Tetrahedron Lett.* **1990**, *31*, 5113. (b) Doyle, M. M.; Jackson, W R.; Perlmutter, P. *Aust. J. Chem.* **1989**, *42*, 1907.

See also: (c) Sato, F.; Ishikawa, H.; Sato, M. *Tetrahedron Lett.* **1981**, *22*, 85. (d) Miller R. B.; McGarvey, G. *J. Org. Chem.* **1978**, *43*, 4424. (e) Miller, R. B.; Reichenbach, T. *Tetrahedron Lett.* **1974**, *15*, 543.

4. (a) Weber, W. P. In *Silicon Reagents for Organic Synthesis*; Springer-Verlag: Berlin, 1983; p. 79. (b) Colvin, E. W. In *Silicon in Organic Synthesis*; Butterworth, Sevenoaks, **1981**; p. 44. (c) Fleming, I.; Dunoguès, J.; Smithers, R. In *Organic Reactions*; Kende, A. S., Ed.; Wiley: New York, 1989; Vol. 37, p. 57. (d) Fleming, I.; Barbero, A.; Walter, D. *Chem. Rev.* **1997**, *97*, 2063.
5. Freifelder, M. *Catalytic Hydrogenation in Organic Synthesis. Procedures and Commentary*; Wiley: New York, 1978; p. 10. Rylander, P. N. *In Selections from the Aldrichimica Acta*; Aldrich: Milwaukee, 1984; p. 195; *Aldrichimica Acta*, **1979**, *12*, 53.
6. March J. *Advanced Organic Chemistry*; 4th ed., Wiley: New York 1992; p. 776.
7. Ref. 4b, p. 48. Ref. 3d. Overman, L. E.; Brown, M. J.; McCann, S. F. *Org. Synth. Coll. Vol. VIII*, **1993**, 609.
8. Harada, K.; Urabe, H.; Sato, F. *Tetrahedron Lett.* **1995**, *36*, 3203.
9. (a) Hamada, T.; Suzuki, D.; Urabe, H.; Sato F. *J. Am. Chem. Soc.* **1999**, *121*, 7342. (b) Launay, V.; Beaudet, I.; Quintard, J.-P. *Synlett* **1997**, 821. (c) Averbuj, C.; Kaftanov, J.; Marek, I. *Synlett* **1999**, 1939. (d) Hungerford, N. L.; Kitching, W. *J. Chem. Soc., Perkin Trans. 1* **1998**, 1839. (e) Hamada, T.; Mizojiri, R.; Urabe, H.; Sato F. *J. Am. Chem. Soc.* **2000**, *122*, 7138. See also: (f) Sawada, D.; Kanai, M.; Shibasaki, M. *J. Am. Chem. Soc.* **2000**, *122*, 10521.
10. Sato, F.; Urabe, H.; Okamoto, S. *Pure Appl. Chem.* **1999**, *71*, 1511. Sato, F.; Urabe, H.; Okamoto, S. *Chem. Rev.* **2000**, *100*, 2835. Sato, F.; Urabe, H.; Okamoto, S. *Synlett* **2000**, 753. See also: Kulinkovich, O. G.; de Meijere, A. *Chem. Rev.* **2000**, *100*, 2789. Eisch, J. J. *J. Organomet. Chem.* **2001**, *617-618*, 148.

11. Buchwald, S. L.; Nielsen, R. B. *Chem. Rev.* **1988**, *88*, 1047. Negishi, E.; Takahashi, T. *Acc. Chem. Res.* **1994**, *27*, 124. Ohff, A.; Pulst, S.; Lefeber, C.; Peulecke, N.; Arndt, P.; Burlakov, V. V.; Rosenthal, U. *Synlett* **1996**, 111. Negishi, E.; Takahashi, T. *Bull. Chem. Soc. Jpn.* **1998**, *71*, 755. Kataoka, Y.; Miyai, J.; Oshima, K.; Takai, K.; Utimoto, K. *J. Org. Chem.* **1992**, *57*, 1973. Hartung, J. B., Jr.; Pedersen, S. F. *J. Am. Chem. Soc.* **1989**, *111*, 5468. Strickler, J. R.; Bruck, M. A.; Wexler, P. A.; Wigley, D. E. *Organometallics* **1990**, *9*, 266.

Appendix

Chemical Abstracts Nomenclature (Collective Index Number); (Registry Number)

1-(Trimethylsilyl)-1-hexyne: Silane, 1-hexynyltrimethyl- (13); 3844-94-8)

Titanium tetraisopropoxide: Titanium, tetrakis(2-propanolato)-, 2-Propanol, titanium (4+) (9); (546-68-9)

Isopropylmagnesium chloride: Magnesium, chloro(1-methylethyl)- (13); (1068-55-9)

(Z)-1-(Trimethylsilyl)-1-hexene: Silane, 1-hexenyltrimethyl-, (Z)- (12); (52835-06-0)

COPPER(I)-CATALYZED PREPARATION OF (E)-3-IODOPROP-2-ENOIC ACID

(2-Propenoic acid, 3-iodo-, (2E)-

HI (57% aq), CuI (0.6%)
reflux, 30 min

Submitted by Darren J. Dixon, Steven V. Ley, and Deborah A. Longbottom.[1]
Checked by Michelle DeRitter and Steven Wolff.

1. Procedure

A 250-mL, three-necked, round-bottomed flask, equipped with a magnetic stirring bar, thermometer and water condenser, is charged with copper(I) iodide (200 mg, 1.1 mmol) (Note 1) and hydriodic acid (57% aq, 40 mL) (Notes 2, 3). Propiolic acid (10 ml, 162 mmol) (Note 4) is added over 1 min via syringe (Note 5), during which time the reaction temperature rises to ~100°C. The reaction mixture is immediately immersed in an oil bath, preheated with a thermostatically controlled stirrer-hotplate to 130°C (Note 6) and a gentle reflux (110°C) is reached after ~4 min. The mixture is heated to reflux for a further 30 min, then the oil bath is removed and replaced with a room temperature water bath. The vigorously stirred reaction mixture is allowed to cool to 28°C over 15 min (Note 7), during which time a large quantity of white needles crystallizes from the reaction mixture (Note 8). The solution is stirred for a further 15 min at room temperature, then filtered through a sintered glass funnel. The crystals are washed with 3 x 70 mL of distilled water and dried with suction for ~1 hr, then over phosphorus pentoxide in a vacuum desiccator to constant weight. Analytically and isomerically pure (E)-3-iodoprop-2-enoic acid [25.3 g (79%)] is obtained as white needles (Note 9).

2. Notes

1. Copper(I) iodide (99%) was purchased from Aldrich Chemical Company, Inc. (Cat. No. 20,554-0) and used as received.

2. Hydriodic acid was purchased from Aldrich Chemical Company, Inc. (Cat. No. 21,002-1) and used as received.

3. Propiolic acid was purchased from Lancaster Synthesis Ltd. (Cat. No. 0669) and used as received.

4. Commercially available HI must be colorless to straw-yellow colored (not brown or darker) for a successful reaction.

5. On larger scales (up to 50 grams of propiolic acid) the addition should still be complete within 1 min to allow the internal reaction temperature to rise to ~100°C.

6. The temperature of the reaction mixture must be raised to the reflux temperature (110°C) immediately after the addition of propiolic acid.

7. On larger scales the water bath may be replaced with an ice bath to reduce the temperature of the reaction mixture to 25-30°C over 15 min.

8. The exact ratio of reagents for this procedure is important for successful formation of pure product in this reaction.

9. Full characterization of the product is as follows: mp 146-149°C; ν_{max} (CH_2Cl_2) cm^{-1}: 3500-2500 (OH, CH), 1701 (C=O); ^{1}H NMR (400 MHz, CD_3OD) δ: 6.90 (1H, d, J =14.9, ICH=C<u>H</u>), 8.07 (1H, d, J =14.9, IC <u>H</u>); ^{13}C NMR (100 MHz, CD_3OD) δ: 103.0 (I<u>C</u>H), 135.7 (ICH=<u>C</u>H), 169.2 (C=O); m/z (EI) 198 (100, $[M]^+$), 181 (11), 153 (12), 127 (53), 71 (15). Anal. Calcd for $C_3H_3IO_2$: C, 18.20; H, 1.52%. Found: C, 18.13; H, 1.49%.

Waste Disposal Information

Aqueous washings from the filtration and isolation of desired product were extracted with ether and disposed of as halogenated waste.

3. Discussion

Pure (E)-3-iodoprop-2-enoic acid can be obtained in one step by heating propiolic acid and HI at 130°C in a sealed tube for extended periods. It can also be formed by thermal isomerization of the pure (Z)-isomer under similar conditions.[2] In a recent improvement to this protocol, (Z)-3-iodoacrylic acid was isomerized by extended periods of heating in hydriodic acid, thus obviating the use of a sealed tube.[3-5]

However, in order to obtain isomerically pure (E)-3-iodoprop-2-enoic acid, a two-step protocol is still required and reactions must be heated for 24 hr in both steps; work-up and purification of the reaction mixture is also necessary in each case. We were hopeful that we could improve upon these procedures by the use of copper(I) salts to catalyze the reaction

Copper(I) halides have been shown previously to accelerate the formation of (Z)-3-iodoprop-2-enoic acid.[6-10] The formation of the (E)-isomer during these low temperature investigations indicated that, at higher temperature, copper catalysis could accelerate the formation of the thermodynamically favored (E)-isomer. Therefore, we decided to investigate the effect of temperature on the copper-catalyzed addition of HI to propiolic acid, with a view to creating a copper-catalyzed, one-step synthesis of the (E)-isomer.

Following optimization studies, we have now found that our method is a dramatic improvement of existing methods and relies on the copper(I)-catalyzed (0.6%) addition-isomerization of hydriodic acid to propiolic acid. The reaction time is short (30 min) and the analytically pure crystalline product is filtered directly from the reaction mixture, providing the desired material in excellent yield (79%). No further purification is necessary, all of which make this procedure the most attractive yet for the synthesis of (E)-3-iodoprop-2-enoic acid.

1. University of Cambridge, Department of Chemistry, Lensfield Road, Cambridge, CB2 1EW, United Kingdom.
2. Abarbri, M.; Parrain, J.-L.; Cintrat J.-C.; Duchene, A. *Synthesis* **1996**, 82.
3. Zoller, T.; Uguen, D. *Tetrahedron Lett.*, **1998**, *39*, 6719.
4. Takeuchi, R.; Tanabe, K.; Tanaka, S. *J. Org. Chem.* **2000**, *65*, 1558.
5. Paterson, I.; Lombart, H.-G.; Allerton, C. *Org. Lett.* **1999**, *1*, 19.
6. Topek, K.; Vsetecka, V.; Procházka, M. *Collect. Czech. Chem. Commun.* **1978**, *43*, 2395.
7. Moss, R. A.; Wilk, B.; Krogh-Jespersen, K.; Westbrook, J. D. *J. Am. Chem. Soc.* **1989**, *111*, 6729.
8. Grandjean, D.; Pale, P.; Chuche, J. *Tetrahedron Lett.* **1992**, *33*, 5355.
9. Grandjean, D.; Pale, P.; Chuche, J. *Tetrahedron* **1993**, *49*, 5225.
10. Keen, S. P.; Weinreb, S. M. *J. Org. Chem.* **1998**, *63*, 6739.

Appendix

Chemical Abstracts Nomenclature (Collective Index Number);

(Registry Number)

(E)-3-Iodoprop-2-enoic acid: 2-Propenoic acid, 3-iodo-, (2E)- (9); (6372-02-7)
Copper(I) Iodide: Copper iodide (CuI) (8, 9); (7681-65-4)
Hydriodic acid (8, 9); (10034-85-2)
Propiolic acid: 2-Propynoic acid (9); (471-25-0)

PREPARATION OF SUBSTITUTED PYRIDINES VIA REGIOCONTROLLED [4 + 2] CYCLOADDITIONS OF OXIMINOSULFONATES: METHYL 5-METHYLPYRIDINE-2-CARBOXYLATE

(2-Pyridinecarboxylic acid, 5-methyl-, methyl ester)

A. 1. $NaNO_2$, $MeOH/H_2O$; 2. TsCl

B. Me_2AlCl, CH_2Cl_2; NaOMe, NCS, THF-MeOH

Submitted by Rick L. Danheiser, Adam R. Renslo, David T. Amos, and Graham T. Wright.[1]

Checked by Helga Krause and Alois Fürstner.

1. Procedure

A. *5-(Tosyloxyimino)-2,2-dimethyl-1,3-dioxane-4,6-dione.* A 100-mL, three-necked, round-bottomed flask equipped with an argon inlet adapter, magnetic stirring bar, rubber septum, and solid addition funnel fitted with a rubber septum (Note 1) is charged with 4.00 g (27.8 mmol) of Meldrum's acid (Note 2) and 23 mL of methanol

(Note 3). To this suspension is added in one portion a solution of 1.91 g (27.8 mmol) of sodium nitrite (Note 4) in 15 mL of water. The reaction mixture is stirred for 2 hr at room temperature to give a deep red solution which is then treated with 4 mL of pH 7 phosphate buffer (Note 5) in one portion and then cooled to 0°C. p-Toluenesulfonyl chloride (4.81 g, 25.2 mmol) (Note 6) is added over 3 min via the solid addition funnel and the cooling bath is removed. The resulting peach-colored mixture is vigorously stirred for 30 min, then the solids are collected by filtration on a Buchner funnel with the aid of 40 mL of cold methanol. The resulting solids are dried at 0.2 mm in a desiccator over phosphorus pentoxide (P_2O_5) for 4 hr to provide 4.85 g (56-59% based on TsCl) of the oximinosulfonate as a white solid, mp 155-156°C (Note 7).

B. *Methyl-5-methylpyridine-2-carboxylate*. A 250-mL, three-necked, round-bottomed flask, equipped with an argon inlet adapter, magnetic stirring bar, rubber septum, and pressure-equalizing addition funnel fitted with a rubber septum (Note 8), is charged with a solution of 4.50 g (13.7 mmol) of oximinosulfonate in 100 mL of dichloromethane (Note 9) and 4.10 mL (2.81 g, 41.1 mmol) of isoprene (Note 10). The solution is cooled at –78°C while 27.5 mL (27.5 mmol) of a 1.0M solution of dimethylaluminum chloride (Me_2AlCl) in hexane (Note 11) is added dropwise via the addition funnel over 9 min. The resulting orange solution is stirred for 3 hr at –78°C to give a yellow solution which is then quenched by the addition of 60 mL of saturated sodium potassium tartrate solution in one portion. The resulting mixture is allowed to warm to 0°C over 30 min, then is transferred to a 500-mL separatory funnel. Dichloromethane (80 mL) and 150 mL of water are added, and the aqueous phase is separated and extracted with three 80-mL portions of dichloromethane. The combined organic phases are washed with 80 mL of saturated sodium chloride solution, dried over magnesium sulfate ($MgSO_4$), filtered, and concentrated at reduced pressure on a rotary evaporator to provide 5.48-5.54 g of cycloadduct as an orange foam which is used in the next step without further purification (Notes 12, 13).

A 250-mL, round-bottomed flask equipped with an argon inlet adapter and magnetic stirring bar (Note 8) is charged with a solution of the crude cycloadduct prepared above in 80 mL of tetrahydrofuran (Note 14) and 80 mL of methanol (Note 3). The solution is cooled at 0°C while 26.3 mL (41.1 mmol) of a 1.56M solution of sodium methoxide (NaOMe) in methanol (Note 15) is added via syringe over 4 min followed by the addition of 1.83 g (13.7 mmol) of N-chlorosuccinimide (Note 16) in one portion. The cooling bath is removed, and the resulting yellow solution is stirred in the dark for 16 hr (Note 17). The reaction mixture is concentrated to a volume of ca. 20 mL by rotary evaporation, then diluted with 150 mL of ethyl acetate and 150 mL of pH 7 phosphate buffer (Note 5). The aqueous phase is separated and extracted with three 100-mL portions of ethyl acetate, and the combined organic phases are extracted with three 100-mL portions of 1.0N HCl. The combined aqueous extracts are neutralized by the slow addition of solid sodium bicarbonate (Note 18), then extracted with three 100-mL portions of ethyl acetate. The combined organic phases are washed with 100 mL of saturated sodium chloride solution, dried over Na_2SO_4, filtered, and concentrated at reduced pressure on a rotary evaporator to afford 1.61-1.76 g (78-85%) of methyl 5-methylpyridine-2-carboxylate as a yellow solid, mp 47-49°C (Notes 19-22).

2. Notes

1. The apparatus is purged with argon and maintained under an atmosphere of argon during the course of the reaction.

2. 2,2-Dimethyl-1,3-dioxane-4,6-dione (Meldrum's acid) was purchased from Aldrich Chemical Co., Inc. and recrystallized before use according to the following procedure. Meldrum's acid (10 g) was dissolved in a minimum amount of warm (40-45°C) acetone (ca. 20 mL). Room temperature water (15 mL) was then added in one portion, resulting in the immediate precipitation of white needles which were

collected by filtration on a Buchner funnel and dried (ca. 0.05 mm) in a desiccator over P_2O_5 overnight.

3. Anhydrous grade methanol was purchased from Mallinkrodt, Inc. and used as received.

4. Sodium nitrite was purchased from Mallinkrodt, Inc. and used as received.

5. Phosphate buffer (pH 7.00) was purchased from VWR Scientific Products, Inc; the checkers used phosphate buffer purchased from Riedel-de Haen.

6. p–Toluenesulfonyl chloride (TsCl) was purchased from Fluka Corporation and recrystallized before use according to the following procedure. TsCl (30 g) was dissolved in a minimum amount (ca. 50 mL) of warm (55-60°C) chloroform to give a yellow solution. To this solution was added 70 ml of warm petroleum ether and 3 g of activated charcoal. After stirring for 10 min, the mixture was filtered and concentrated to half volume. Crystals of TsCl precipitated and after 30 min were collected by filtration on a Buchner funnel. Drying at 0.05 mm for 1 hr afforded 22 g of TsCl as white crystals, mp 67-68°C.

7. The oximinosulfonate has the following spectroscopic properties: IR ($CHCl_3$) cm^{-1}: 3020, 1790, 1765, 1596, 1400, 1290; ^1H NMR (300 MHz, $CDCl_3$) δ: 1.78 (s, 6 H), 2.46 (s, 3 H), 7.40 (d,2 H, J = 8.4), 7.93 (d, 2 H, J = 8.4); ^{13}C NMR (75 MHz, $CDCl_3$) δ: 21.8, 28.1, 106.8, 129.5, 130.1, 130.2, 138.8, 147.1, 149.6, 154.7. Anal. Calcd for $C_{13}H_{13}NO_7S$: C, 47.70; H, 4.00; N, 4.28; Found: C, 47.76; H, 4.02; N, 4.22.

8. The apparatus is flame-dried under reduced pressure and then maintained under an atmosphere of argon during the course of the reaction.

9. Dichloromethane was purchased from J.T. Baker, Inc. and purified by pressure filtration through activated alumina.

10. Isoprene was purchased from Aldrich Chemical Co., Inc. and distilled under argon at atmospheric pressure immediately before use.

11. Dimethylaluminum chloride in hexanes (1.0M) was purchased from Aldrich Chemical Company, Inc.

12. A pure sample of the intermediate cycloadduct can be obtained by column chromatography on silica gel (elution with 1% methanol/dichloromethane). The cycloadduct exhibits the following spectroscopic properties: IR ($CHCl_3$) cm^{-1}: 3020, 1780, 1750, 1385, 1300; ^1H NMR (300 MHz, $CDCl_3$) δ: 1.67 (s, 3 H), 1.69 (s, 3 H), 1.88 (s, 3 H), 2.48 (s, 3 H), 2.71 (br dd, 2 H, J = 1.2, 3.3), 3.93 (s, 2 H), 5.33 (br., 1H), 7.36 (d, 2H, J = 7.9), 7.81 (d, 2 H, J = 8.4); ^{13}C NMR (75 MHz, $CDCl_3$) δ: 20.3, 21.7, 28.5, 29.3, 32.8, 57.4, 66.3, 106.2 , 113.9, 129.2, 129.55, 129.62, 131.2, 145.9, 164.0.

13. *CAUTION*: The checkers experienced a rather vigorous decomposition of this product upon evaporating the solvent in the rotary evaporator and drying of the crude material at 39°C (bath temperature) and 15 mm. No hazards, however, were encountered when this operation was carried out at ambient temperature (22°C) at 38 mm. The resulting crude material was directly used in the next step.

14. Tetrahydrofuran was purchased from J.T. Baker, Inc. and purified by pressure filtration through activated alumina.

15. A 1.56M solution of sodium methoxide was prepared by careful addition of 2.69 g of sodium metal (cut into ca. 15 pieces) to 75 mL of methanol at 0°C under an atmosphere of argon.

16. N-Chlorosuccinimide (NCS) was purchased from Aldrich Chemical Company, Inc. and recrystallized before use according to the following procedure. NCS (25 g) was dissolved in 125 mL of glacial acetic acid at 60-65°C in a 250-mL, Erlenmeyer flask and the resulting solution was allowed to cool to room temperature. The Erlenmeyer flask was then placed in an ice-water bath (15-20°C), and white flakes of NCS immediately precipitated and were collected by filtration on a Buchner funnel (washing with 20-mL of glacial acetic acid and then with two 20-mL portions of hexanes) and then dried at 0.5 mm for 6 hr and stored in the dark until use.

17. For this purpose, the flask was wrapped in aluminum foil.

18. Sodium bicarbonate was added until gas evolution was no longer observed.

19. The purity of this material was confirmed by spectroscopic and elemental analysis: IR (KBr) cm^{-1}: 3091, 3048, 3004, 2956, 1730, 1591, 1448, 1320, 1251, 1126, 1032, 781, 704; ^{1}H NMR (300 MHz, $CDCl_3$) δ: 2.43 (s, 3 H), 4.00 (s, 3H), 7.64 (dd, 1 H, J = 7.8, 2.0), 8.05 (d, 1 H, J = 8.3), 8.57 (d, 1 H, J = 2.0); ^{13}C NMR (75 MHz, $CDCl_3$) δ: 18.5, 52.6, 124.6, 137.1, 137.3, 145.2, 150.2, 165.6. Anal. Calcd for $C_8H_9NO_2$: C, 63.56; H, 6.00; N, 9.27; Found: C, 63.21; H, 6.02; N, 9.43; TLC: R_f = 0.11 (elution with 25% ethyl acetate-hexanes; phosphomolybdic acid in ethanol stain).

20. If desired, further purification can be effected by chromatography (Note 21) or, alternatively, by recrystallization according to the following procedure: the crude product is dissolved in 20 mL of warm (50°C) hexanes and the solution is decanted from an insoluble orange solid and transferred to a 50-mL, round-bottomed flask. Concentration provides 1.34 g (65%) of methyl 5-methylpyridine-2-carboxylate as a colorless solid, mp 53-54°C.

21. Purification by column chromatography was carried out according to the following procedure: silica gel (10 g) is added to the dried solution of crude product obtained from the work up, and the solution is concentrated by rotary evaporation to afford a free-flowing powder which is placed at the top of a column (4.5 cm diameter) of 40 g of silica gel (230-400 mesh) and eluted with 25% ethyl acetate/hexane (containing 1% triethylamine) to afford 1.47 g (71%) of methyl 5-methylpyridine-2-carboxylate as a pale yellow solid, mp 54-55°C.

22. The checkers obtained the crude material as a yellow-brown syrup (1.86 g, 89%). Column chromatography as described in Note 21 afforded 1.29 g (63%) of the title compound as a pale yellow solid, mp 52-53°C.

Waste Disposal Information

All toxic materials were disposed of in accordance with "Prudent Practices in the Laboratory"; National Academy Press; Washington, DC, 1995.

3. Discussion

The structures of a number of natural products incorporate the pyridine ring as do important commercial compounds including herbicides, insecticides, fungicides, and a variety of medicinal agents. Among the numerous strategies that have been developed for pyridine synthesis,[2] methods based on the Diels-Alder reaction are particularly attractive due to their intrinsic convergent character and regiochemical features. Although a variety of azadienophiles[3] have been used for the synthesis of tetrahydropyridines, few examples of their application in the synthesis of pyridines have previously been reported. The current procedure illustrates a versatile method for the synthesis of substituted pyridines via [4+2] cycloaddition of an oximinosulfonate derived from Meldrum's acid with conjugated dienes. Diels-Alder reactions of this new azadienophile are subject to Lewis acid promotion, permitting highly regioselective cycloadditions with a wide range of diene substrates under relatively mild reaction conditions.

As described here, we have developed a simple procedure that delivers multigram quantities of the key oximinosulfonate in good yield and excellent purity in one operation without the need for additional purification.[4] The oximinosulfonate is an easily handled solid that can be transferred in air and is stable for months when stored under argon in a refrigerator. Reaction of the oximinosulfonate with a conjugated diene and 2 eq of Me_2AlCl leads to an efficient Diels-Alder reaction, providing cycloadducts which can then be transformed to substituted pyridines in a single synthetic operation.

Significantly, the regiochemical course of the cycloaddition is opposite to that observed with conventional imino dienophiles, thus producing heterocycles with substitution patterns that cannot be accessed using prior imino-dienophile Diels-Alder methodology. A series of experiments in which the Me_2AlCl stoichiometry was varied from 0.1-2.5 equivalents established the requirement that a full 2.0 eq of the Lewis acid be employed. As shown below, it is likely that the second eq of Me_2AlCl serves to promote the ionization of chloride from an initial 1:1 Lewis acid complex and thereby generate the more reactive ionic 2:1 complex.[5,6]

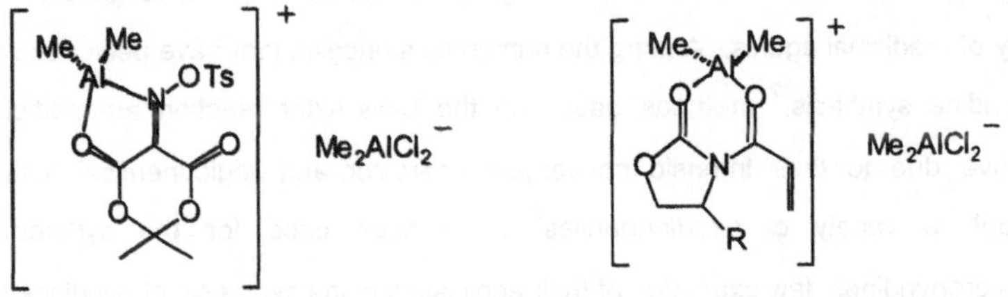

The cycloadducts obtained in the oximinosulfonate Diels-Alder reaction are best converted directly to pyridines without purification. Exposure of the spiro-fused cycloadducts to a combination of NCS and sodium methoxide brings about cleavage of the dioxanedione ring with concomitant elimination of acetone and carbon dioxide. Elimination of tosylate from the resulting ester enolate then generates a dihydropyridine, and subsequent chlorination by NCS and elimination of HCl finally provides the desired aromatic pyridine product.

The Lewis acid-promoted reaction of our oximinosulfonate with dienes and the conversion of the resulting cycloadducts to pyridines comprises a new annulation method for the synthesis of substituted pyridines from conjugated dienes. As illustrated in the Table, very good overall yields are obtained in reactions of 2-substituted dienes, providing 5-substituted pyridine-2-carboxylates. Reactions with 1-substituted dienes yield 3-substituted pyridines, and disubstituted dienes react smoothly to afford trisubstituted pyridines in good yield. Polycyclic systems are obtained when dienes such as 1-vinyl-1-cyclohexene are employed in the annulation.

1. Department of Chemistry, Massachusetts Institute of Technology, Cambridge, MA 02139. We thank the National Institutes of Health (GM 28273) for generous financial support. D.T.A. was supported in part by NIH training grant CA 09112.
2. For general reviews of the synthesis and chemistry of pyridines, see: (a) Scriven, E. F. V. In *Comprehensive Heterocyclic Chemistry*; Katritzky, A. R., Rees, C. W., Eds.; Pergamon: New York, 1984; Vol. 2, Part 2A, pp. 165-314. (b) Yates, F. S. In *Comprehensive Heterocyclic Chemistry*; Katritzky, A. R., Rees, C. W., Eds.; Pergamon: New York, 1984; Vol. 2, Part 2A, pp. 511-524. (c) Tomasik, P.; Ratajewicz, Z. In *Pyridine-Metal Complexes*; Newkome, G. R., Strekowski, L., Eds.; The Chemistry of Heterocyclic Compounds; Wiley: New York, 1985; Vol. 14, Part 6A. (d) Jones, G. In *Comprehensive Heterocyclic Chemistry*; Katritzky, A. R., Rees, C. W., Eds.; Pergamon: New York, 1984; Vol. 2, Part 2A, pp. 395-510. (e) *Pyridine and its Derivatives*; Newkome, G. R., Ed; The Chemistry of Heterocyclic Compounds; Wiley: New York, 1984; Vol. 14, Part 5.
3. For reviews of the application of azadienophiles in the Diels-Alder reaction, see: (a) Weinreb, S. M. In *Comprehensive Organic Synthesis*; Trost, B. M.; Fleming, I., Eds.; Pergamon Press: Oxford, 1991, Vol. 5, pp. 401-449. (b) Weinreb, S. M.; Levin, J. I. *Heterocycles* **1979**, *12*, 949. (c) Weinreb, S. M.; Staib, R. R. *Tetrahedron* **1982**, *38*, 3087. (d) Boger, D. L.; Weinreb, S. M. *Hetero Diels-Alder Methodology in Organic Synthesis*; Academic: San Diego, 1987; Chapter 2. (e) Tietze, L. F.; Kettschau, G. *Top. Curr. Chem*. **1997**, *189*, 1.

4. Renslo, A. R.; Danheiser, R. L. *J. Org. Chem.* **1998**, *63*, 7840.
5. Lehmkuhl, H.; Kobs, H.-D. *Liebigs Ann. Chem.* **1968**, *719*, 11.
6. Evans, D. A.; Chapman, K. T.; Bisaha, J. *J. Am. Chem. Soc.* **1988**, *110*, 1238.

Appendix

Chemical Abstracts Nomenclature (Collective Index Number); (Registry Number)

Methyl 5-methylpyridine-2-carboxylate: 2-Pyridinecarboxylic acid, 5-methyl-, methyl ester (9); (260998-85-4)

5-(Tosyloxyimino)-2,2-dimethyl-1,3-dioxane-4,6-dione: 1,3-Dioxane-4,5,6-trione, 2,2-dimethyl-, 5-O-[(4-methylphenyl)sulfonyl]oxime (9); (215436-24-1)

Meldrum's acid: 1,3-Dioxane-4,6-dione, 2,2-dimethyl- (9); (2033-24-1)

Sodium nitrite: Nitrous acid, sodium salt (8, 9); (7632-00-0)

p-Toluenesulfonyl chloride: Benzenesulfonyl chloride, 4-methyl- (9); (98-59-9)

Isoprene: 1,3-Butadiene, 2-methyl- (9); (78-79-5)

Dimethylaluminum chloride: Aluminum, chlorodimethyl- (8, 9); (1184-58-3)

N-Chlorosuccinimide: 2,5-Pyrrolidinedione, 1-chloro- (9); (128-09-6)

TABLE

[4+2] Pyridine Annulations[a]

diene[b]	pyridine	% yield[c]
	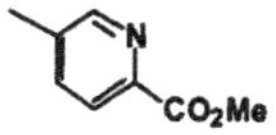 CO_2Me	78-85
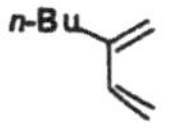 *n*-Bu	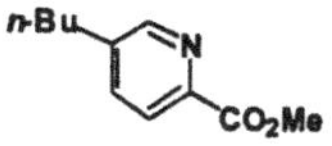 *n*-Bu, CO_2Me	72
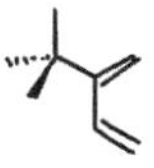	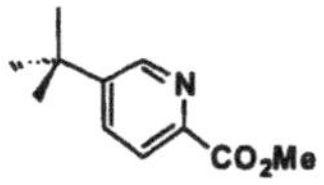 CO_2Me	73
	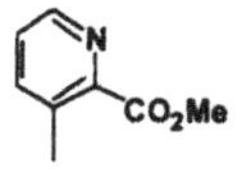 CO_2Me	56-57
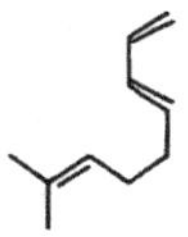	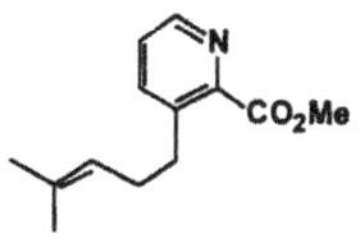 CO_2Me	60
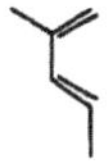	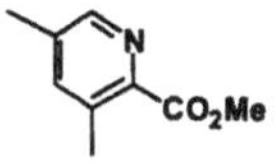 CO_2Me	70
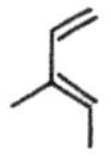	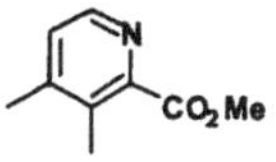 CO_2Me	40
	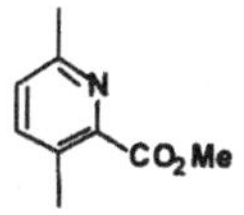 CO_2Me	40[d]
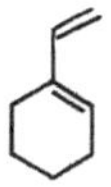	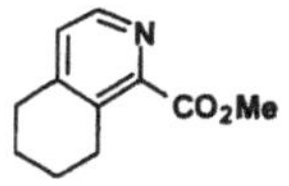 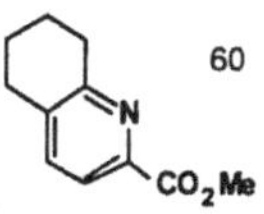 CO_2Me CO_2Me 83 : 17	60

[a] Conditions: (a) oximinosulfonate, 2.0 equiv Me_2AlCl, CH_2Cl_2, -78 °C, 2-4 h; (b) 3.0 equiv MeONa, 1.0 equiv NCS, MeOH-THF (1:1), rt, 14-16 h. [b] 1.5-3.0 equiv of diene was employed. [c] Isolated overall yield for two steps. [d] 5.0 equiv of NaOMe and 4.0 equiv of NCS were used.

SYNTHESIS AND [3+2] CYCLOADDITION OF A 2,2-DIALKOXY-1-METHYLENECYCLOPROPANE: 6,6-DIMETHYL-1-METHYLENE-4,8-DIOXASPIRO[2.5]OCTANE and cis-5-(5,5-DIMETHYL-1,3-DIOXAN-2-YLIDENE)HEXAHYDRO-1(2H)-PENTALEN-2-ONE

(4,8-Dioxaspiro[2.5]octane, 6,6-dimethyl-1-methylene-)

Submitted by Masaharu Nakamura,[1] Xiao Qun Wang,[1] Masahiko Isaka,[1] Shigeru Yamago,[2] and Eiichi Nakamura.[1]

Checked by Jimmy Wu and Edward J. J. Grabowski.

1. Procedure

Caution! All operations should be performed in a well-ventilated hood, and care should be taken to avoid skin contact with 1,3-dichloroacetone and cyclopropenes.

A. *2,2-Bis-(chloromethyl)-5,5-dimethyl-1,3-dioxane* (**1**). A mixture of 1,3-dichloroacetone (152 g, 1.20 mol, an irritant), neopentyl glycol (138 g, 1.32 mol), p-toluenesulfonic acid (4.6 g, 0.024 mol) and benzene (100 mL) (Note 1) is heated to reflux for 19 hr in a 500-mL, round-bottomed flask equipped with a Dean-Stark trap and a condenser with azeotropic removal of water. After separation of water ceases, the reaction mixture is partitioned between hexane (500 mL) and saturated sodium bicarbonate ($NaHCO_3$) (200 mL). The organic phase is washed with water (100 mL) and saturated sodium chloride (NaCl) (100 mL), dried over $MgSO_4$, and concentrated on a rotary evaporator. Distillation of the residue yields, after about 5 g of forerun, 249 g (97%) of the acetal **1** as a colorless oil (bp 99-100°C, 3.5 mm) (Note 2).

B. *1,6,6-Trimethyl-4,8-dioxaspiro[2.5]oct-1-ene* (**3**) (Note 3). A solution of sodium amide in liquid ammonia is prepared according to a procedure previously described (Notes 4, 5). An oven-dried, 2-L, three-necked, round-bottomed flask is equipped with a mechanical stirrer, nitrogen gas inlet, and dry ice/acetone condenser protected with a drying tube containing potassium hydroxide pellets. The flask is flushed with nitrogen introduced through the gas inlet, then is placed in a dry ice/acetone bath. The nitrogen source is replaced with a hose connected to a cylinder of ammonia (NH_3). Gaseous NH_3 is introduced to the flask condensing ca. 400 mL of NH_3 and gentle stirring is started. The NH_3 inlet is replaced with a glass stopper and the dry ice/acetone bath is replaced with a –35°C bath (electronically controlled or a dry ice/trichloroethylene bath). Crystals of $Fe(NO_3)_3 \cdot 9H_2O$ (0.3 g) are added through the neck having a stopper. To the resulting orange solution is added a small (about 5 mm^3) cube of sodium. The solution is stirred (Note 6) until the blue color disappears and fine black particles appear. Pieces of sodium

(21.44 g, 0.930 mol) are then added over 35 min. After 20 min, the solution turns into a dark gray suspension with a white precipitate. The cooling bath is replaced with a dry ice/acetone bath, and the gas inlet is replaced with a pressure-equalizing addition funnel containing a solution of 2,2-bis-(chloromethyl)-5,5-dimethyl-1,3-dioxane **1** (63.93 g, 0.300 mol) in 150 mL of dry ether (Et_2O). This solution is added dropwise to the slurry of sodium amide in liquid NH_3 over 1 hr. The addition funnel is rinsed with 20 mL of dry Et_2O. The cooling bath is removed, and the mixture is stirred for 3 hr (Note 7), then is cooled again with a dry ice/acetone bath. After 10 min, a solution of freshly distilled methyl iodide (44.71 g, 0.315 mol) in 80 mL of dry Et_2O is added via the addition funnel over 1 hr (Notes 8, 9); the funnel is rinsed with 20 mL of dry Et_2O. After stirring for 15 min, the cooling bath is removed and the mixture is stirred for 1 hr. The mixture is cooled with a dry ice/acetone bath again and solid ammonium chloride (NH_4Cl) (20.24 g, 0.378 mol) is added in several portions over 5 min. The dry ice condenser is removed and the ammonia is allowed to evaporate. The cooling bath is replaced with a water bath (ca. 30°C), and a 1:1 mixture of dry Et_2O and dry pentane (400 mL) is added through the addition funnel over 10 min with vigorous stirring. The water bath temperature is maintained between 25-30°C. After evaporation of most of the ammonia (1.5 to 2 hr), the ethereal solution is filtered by suction through a pad of Hyflo Super Cel to remove the inorganic salts. The filter cake is washed three times with 80 mL of Et_2O. The combined filtrate and washes are concentrated under reduced pressure (30-40 mm, 25°C) (Note 10) and the residue is distilled to yield **3** as a colorless oil (32.3 g, 70%; bp 58-61°C, 6-7 mm) (Notes 11-14). While this material is adequate for use in Step C, a pure sample can be obtained by flash chromatography of the crude product followed by distillation. For 36.1 g of crude product (obtained on a 0.25 mol scale), chromatography is performed with 290 g of silica gel (Merck, Kieselgel 60) and Et_2O:pentane (5:95) as eluent, collecting 160-mL fractions. Fractions 3-16 containing the product are combined. Removal of the solvent under reduced pressure followed by vacuum distillation afforded 29.3 g of **3** in 76% yield (99% pure by GC).

C. *6,6-Dimethyl-1-methylene-4,8-dioxaspiro[2.5]octane* (**5**). An oven-dried, 100-mL, round-bottomed flask, equipped with a magnetic stirring bar and three-way stopcock, which is connected to a vacuum/nitrogen line, is flushed with nitrogen. The flask is placed in a plastic bag filled with nitrogen, and powdered potassium tert-butoxide (1.32 g, 11.8 mmol) (Note 15) is introduced quickly. The flask is connected to a nitrogen source, tert-butyl alcohol (7.11 g, 96 mmol) and 40 mL of Et_2O (Note 16) are introduced by syringe, and the solution is stirred for 10 min at room temperature.

An oven-dried 300-mL, round-bottomed flask, equipped with a magnetic stirring bar and three-way stopcock, connected to a vacuum/nitrogen line, is charged with 1,6,6-trimethyl-4,8-dioxaspiro[2.5]oct-1-ene (46.3 g, 0.30 mol) under nitrogen. Ether (100 mL) is introduced via syringe, and the solution is stirred with external cooling at –78°C (dry ice/acetone bath). The solution of potassium tert-butoxide is added to this solution through a 1.5 mm i.d. cannula over 5 min. After stirring for 5 min at this temperature, the cooling bath is removed and the solution is allowed to warm to room temperature (Note 17). After stirring for 4 hr (Note 18), 15 mL of 1N HCl is added in one portion. The mixture is transferred to a separatory funnel with the aid of 30 mL of Et_2O, and the aqueous phase is separated. The organic phase is washed successively with 5 mL each of saturated $NaHCO_3$ and saturated NaCl, and is dried over Na_2SO_4. After concentration on a rotary evaporator, a pale yellow oil is obtained. Distillation under reduced pressure (Note 17) (57-59°C/7.3 mm) affords 36.49 g (79% yield) of **5** (Notes 19, 20) after ca. 1.5 g of a forerun. The material is sufficiently pure for the subsequent step.

D. *cis-5-(5,5-Dimethyl-1,3-dioxan-2-ylidene)hexahydro-1(2H)-pentalen-2-one* (**6**). An oven-dried, 50-mL, round-bottomed flask, equipped with a magnetic stirring bar and three-way stopcock, is flushed with nitrogen. A mixture of 6,6-dimethyl-1-methylene-4,8-dioxaspiro[2.5]octane, **5**, (8.48 g, 55 mmol) and 2-cyclopenten-1-one (4.52 g, 55 mmol) in 15 mL of acetonitrile is introduced via syringe and the solution is heated at 60°C for 12 hr. The three way stopcock is replaced with a distillation head. Solvent is removed by distillation (ca. 30-120°C/ca. 20-1.4 mm) and the residue is distilled under reduced pressure (142-143°C, 1.4 mm) to afford **6** (10.0 g, 48 mmol, 77%), which crystallizes upon standing at room temperature (Notes 21, 22).

2. Notes

1. The following chemicals were purchased from Kanto Chemical Co. Inc. and used as received: neopentyl glycol, p-toluenesulfonic acid, sodium, and n-BuLi. N,N,N',N'-Tetramethylethylenediamine was purchased from Aldrich Chemical Company, Inc. and distilled before use. 1,3-Dichloroacetone was obtained from Wacker Chemicals East Asia and used as received. Reagent grade benzene, pentane, ether, THF, tert-butyl alcohol, acetonitrile, and toluene were purchased from Wako Chemicals Industries Ltd. Benzene, pentane, and tert-butyl alcohol were distilled from CaH_2; ether and THF from sodium benzophenone ketyl immediately before use; acetonitrile successively from P_2O_5 and anhydrous K_2CO_3; and toluene from $LiAlH_4$. Potassium tert-butoxide and 2-cyclopenten-1-one were purchased from Tokyo Kasei Kogyo Co. Ltd.; the ketone was distilled before use.

2. The product was pure by 270 MHz ^1H NMR analysis. Spectral properties: IR (neat) cm^{-1}: 2950, 2860, 1105, 1025, 770; ^1H NMR (270 MHz, $CDCl_3$) δ: 1.00 (s, 6 H), 3.57 (s, 4 H), 3.80 (s, 4 H).

3. Important notes: (1) Since the reactions are sensitive to temperature effects, care must be taken to allow the reaction mixture to reach thermal equilibrium with the cooling bath at each stage of the operation. (2) Because of the rather unstable nature of the product in its impure form, the entire procedure except the final distillation must be carried out within a day. This will take at least 12 hr. See also Note 10.

4. One of the procedures for converting sodium to sodium amide described in *Organic Syntheses* is used.[3]

5. The use of commercial $NaNH_2$ is not recommended since the yield and the purity of the cyclization product are heavily dependent on the purity of $NaNH_2$.

6. Gentle stirring (e.g., 120 rpm) is recommended to avoid loss of the amide base which may adhere to the upper part of the flask. This was found to be a good practice to achieve the most accurate base/substrate molar ratio.

7. The reaction may be complete in about 2 hr. An aliquot may be removed and analyzed by GC for the intermediate 2-chlorocyclopropanone acetal. In case of a deficiency of sodium amide, a small amount of the chloride may remain but can be removed by distillation. The analysis may be performed on a 0.25 mm i.d. x 25 m capillary GC column (HR-1, Shinwa Chemical Industries, Ltd.) at 80°C. Typical retention times for the chlorocyclopropanone acetal and **3** are 9.9 and 4.6 min, respectively. The amount of the former may be estimated by comparison of its peak area with that of **3** assuming equal detector response factors of the two compounds.

8. Rapid addition of methyl iodide causes the formation of a significant amount of 2,3-dialkylated compound and 2-methylenecyclopropanone acetal (by isomerization of the olefin).

9. If methyl iodide is replaced with solid NH_4Cl (3~4 eq to **1**, added very carefully in several portions), the unsubstituted cyclopropenone acetal **4** can be obtained in the same manner. The submitters carried out this reaction on a 0.3-1 mol scale in 70-85% distilled yield.

10. Care must be taken not to lose material upon concentration. In case of a lack of time, the crude product may be kept under nitrogen in a freezer (–20°C) and distilled later.

11. In order to obtain a good yield, it is important to carry out the distillation as fast as possible (<25 min), but with the bath temperature below 100°C.

12. This product may contain 6~8% of the unsubstituted cyclopropenone acetal **4** and 3~5% of 2,3-dimethylcyclopropenone acetal and ca. 1% of methylene-cyclopropanone acetal **5** as determined by capillary GC analysis on a 0.25 mm i.d. x 25 m capillary column

(HR-1) at 80°C. Typical retention times for **3**, **4**, **5**, and 2,3-dimethyl cyclopropenone acetal are 6.4, 4.3, 5.8 and 9.6 min, respectively. The checkers used material of this purity for Step C.

13. The product has the following physical properties: IR (neat) cm^{-1}: 2950, 1840 (w), 1735, 1280, 1070; ^{1}H NMR (270 MHz, $CDCl_3$) δ: 0.99 (s, 3 H, C$\underline{H}_3$), 1.07 (s, 3 H, C$\underline{H}_3$), 2.19 (3 H, cyclopropyl CH_3), 3.63 (s, 4 H, OC$\underline{H}_2$C), 7.34 (s, 1 H, C=C$\underline{H}$); ^{13}C NMR (100 MHz, $CDCl_3$) δ: 10.27, 22.14, 22.45, 30.47, 77.10, (2C), 83.24, 116.04, 125.42.

14. The cyclization in Step B is an improvement of Butler's procedure for the synthesis of **4**,[4,5] which employs less convenient reagents, KNH_2 and 1-bromo-3-chloroacetone acetal. Beside the acetals derived from neopentyl glycol, those derived from ethanol, 1,3-propanediol and 2,4-pentanediol have been synthesized by the present method.[7a] The second part of Step B involves the formation and the electrophilic trapping of cyclopropenyl anion **2**, which is the key element of the present preparations. Step B provides a simple route to substituted cyclopropenones, but the reaction is limited to alkylation with alkyl halides. The use of lithiated and zincated cyclopropenone acetal,[7] on the other hand, is more general and permits the reaction with a variety of electrophiles; alkyl, aryl and vinyl halides, Me_3SiCl, Bu_3SnCl, aldehydes, ketones, and epoxides. Repetition of the lithiation/alkylation sequence provides disubstituted cyclopropenone acetals.

15. High purity of potassium tert-butoxide is crucial for a clean reaction. The tert-butanol must be anhydrous.

16. A 1:1 mixture of dry dimethylsulfoxide/Et_2O was successfully used for other higher homologues of alkylidenecyclopropanone acetals.

17. The reaction can be monitored either by TLC or GC. TLC analysis may be performed with a plate (Merck No. 5765), eluting with a 1:9 mixture of ethyl acetate and hexane. The R_f values of **3** and **5** are 0.22 and 0.44, respectively. GC analysis performed

under the same conditions as described in Note 12 separates these two compounds, which have retention times of 6.4 and 5.8 min, respectively.

18. In the event that the reaction does not start or has not been completed, another 0.6-0.9 g portion of potassium tert-butoxide dissolved in t-BuOH/Et_2O may be added.

19. It is important to keep the bath temperature as low as possible. Distillation above 100°C would result in greatly reduced yield due to thermal decomposition of **5**.

20. The product has the following physical properties: IR (neat) cm^{-1}: 3150 (w), 1270, 1255, 1245, 1150, 1140, 1070, 1030, 1015, 970, 905; ^1H NMR (200 MHz, $CDCl_3$) δ: 1.04 (s, 3 H, C$\underline{H}_3$), 1.06 (s, 3 H, C$\underline{H}_3$), 1.62 (dd, 2 H, J = 3.1, 2.4, cyclopropyl C$\underline{H}_2$), 3.63 (s, 4 H, OC$\underline{H}_2$C), 5.45 (t, 1 H, J = 2.4, C=CH$\underline{H}$), 5.81 (t, 1 H, J = 3.1, C=CH$\underline{H}$); Anal. Calcd for $C_9H_{14}O_2$: C, 70.10; H, 9.15. Found: C, 70.02; H, 9.25.

21. The cycloadduct has the following spectral properties: ^1H NMR (500 MHz, CD_3CN) δ: 0.96 (s, 3 H), 0.99 (s, 3 H), 1.64-1.70 (m, 1 H), 2.00 (br dd, 1 H, J = 15.6, 4.6), 2.02-2.10 (m, 1 H), 2.21 (distorted t, 2 H, J = 7.6), 2.33 (br dd, 1 H, J = 16.5, 3.7), 2.39-2.49 (m, 2 H), 2.52 (ddd, 1 H, J = 9.6, 4.7, 2.1) 2.74-2.81 (m, 1 H), 3.59 (d, 1 H, J = 7.3), 3.60 (d, 1 H, J = 7.3), 3.61 (s, 2 H); ^{13}C NMR (125 MHz, CD_3CN) δ: 22.13 (q), 22.18 (q), 26.57 (t), 29.80 (t), 31.26 (s), 33.19 (t), 37.46 (t), 42.21 (d), 53.25 (d), 77.42 (t), 77.44 (t), 97.92 (s), 148.69 (s), 222.44 (s).

The ketene acetal functionality of the cycloadduct is moisture sensitive. It can be readily hydrolyzed to the cyclopentanone ester **7** upon treatment with aqueous acetic acid. Spectral properties of **7** are as follows: the major, more polar product–IR (neat) cm^{-1}: 3440, 1730; ^1H NMR (200 MHz, $CDCl_3$) 0.91 (s, 6 H), 1.60 (dt, 1 H, J = 13.3, 9.7), 1.74-2.49 (m, 8 H), 2.61 (dt, 1 H, J = 9.5, 5.9), 2.71-2.99 (m, 2 H), 3.29 (br s, 2 H), 3.92 (s, 2 H); Anal. Calcd for $C_{14}H_{22}O_4$: C, 66.11; H, 8.72. Found: C, 66.00; H, 8.72; the minor, less polar product–IR (neat) cm^{-1}: 3440, 1730, 1720; ^1H NMR (200 MHz, $CDCl_3$) δ: 0.92 (s, 6 H), 1.49-1.86 (m, 2 H), 2.03-2.37 (m, 7 H), 2.57-2.84 (m, 2 H), 2.85-3.06 (m, 1 H), 3.29 (br d, 2

H, J = 2.3), 3.94 (s, 2 H); Anal. Calcd for $C_{14}H_{22}O_4$: C, 66.11; H, 8.72. Found: C, 66.00; H, 8.86.

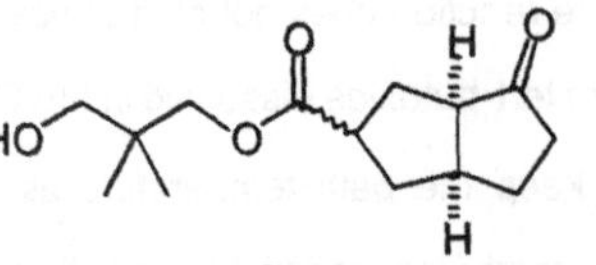

7

22. This product may contain ca. 5% of impurities as judged by capillary GC analysis on a 0.25 mm i.d. x 25 m capillary column (HR-1) at 170°C. Typical retention times for the title compound and the side product are 8.8 and 4.5 min respectively.

3. Discussion

The procedures described herein illustrate the preparation of a substituted cyclopropenone acetal and an alkylidene cyclopropanone acetal. The latter compound has been used to generate a dipolar trimethylenemethane (TMM) species that undergoes [3+2] cycloaddition with electron-deficient 2p-electron C=C and C=X compounds.[6]

The substituted cyclopropenone acetal synthesized in Step B can be readily hydrolyzed to the corresponding cyclopropenone.[7] This synthetic sequence provides the best and most versatile current synthetic route to substituted cyclopropenones.[8] The deficiencies of conventional procedures are precisely the synthesis of cyclopropenones with aliphatic substituents and functional groups, for which the present method has proven to be particularly useful.[9]

Cyclopropenones show considerable biological activity,[10,11] and have recently been employed as a key structural unit for a novel inhibitor of a cysteine protease.[12] The utility of cyclopropenone acetals has recently been recognized for vinylcarbene formation,[13,14] asymmetric synthesis,[15] and other processes.[16] Cycloaddition reactions of cyclopropenone acetals and congeners have also proven to be useful for chiral functionalization of buckminsterfullerenes.[17]

Step C describes a synthesis of the 1,1-dialkoxy-2-methylenecyclopropane (**5**) from 1,1-dialkoxy-2-methylenecyclopropene (**3**).[18] As described in Step D, upon thermolysis under mild conditions, **5** undergoes a [3+2] cycloaddition reaction by the reversible generation of a reactive dipolar TMM **8** (eq 1).[19] Various 2-alkylidene-1,1-dialkoxycyclopropanes can also be prepared by the same procedure from the corresponding 2-alkyl-1,1-dialkoxycyclopropenes, and have been shown to undergo regio- and stereoselective [3+2] cycloaddition reactions.[20] The use of a nitrogen atmosphere and dry solvent is not essential, and the reactions can be carried out under air in the presence of a small amount of water (the use of a slight excess of the methylenecyclopropane removes any water without affecting the yield of the cycloaddition product). Solvent is also not essential; the reaction rate and course may be affected by the choice of solvent (or absence of solvent).

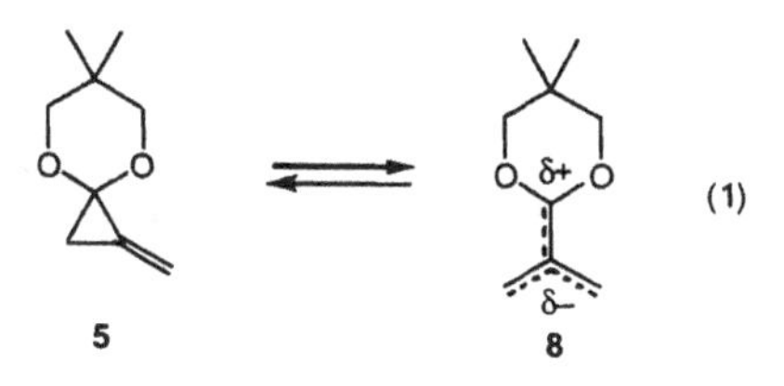

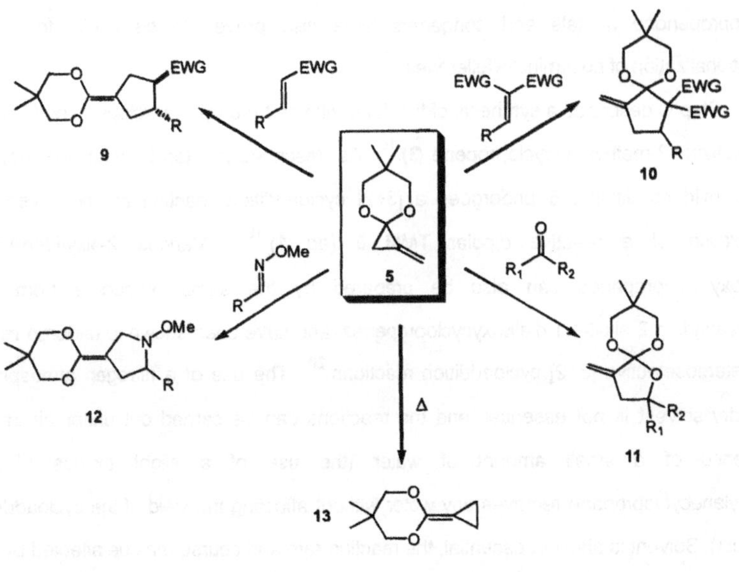

Olefins substituted with a single ester, nitrile, or ketone group function as good TMM acceptors to afford ketene acetals **9** (Table 1, entries 1-4). The cycloaddition reaction takes place stereoselectively with retention of the olefin geometry and "endo" orientation of the directing groups.[20] In contrast, the reaction with olefins whose reduction potential is larger than –1.8 eV (vs. SCE) gives mainly the acetal of an α-methylenecyclopentanone **10** via a single electron transfer process.[21] This cycloaddition can be used for the functionalization of buckminsterfullerene (C_{60} and C_{70})[22,23] whose reduction potential is –0.42 eV (see below). A radioactive version of **2** has been prepared by using $^{14}CH_3I$ in Step B and has been employed for pharmacokinetic studies of water soluble C_{60}.[24] Cycloaddition to electron-deficient acetylenes takes place smoothly to give cyclopentene carboxylic esters in one step (entries 7 and 8).[25] Cycloaddition to carbonyl compounds gives tetrahydrofuran derivatives (entries 9-11),[26] and to an O-alkyloxime affords a pyrrolidine (entry 12).[27] Prolonged heating gives a highly reactive ketene acetal **13**, which serves as a useful synthon of cyclopropanecarboxylate enolate.[28]

C_{60} → 27% + 34%

82%, 120-160°C

1. Department of Chemistry, The University of Tokyo, Bunkyo-ku, Hongo, Tokyo 113, Japan.
2. Department of Synthetic Chemistry & Biological Chemistry, Kyoto University, Sakyo-ku, Kyoto 606-01, Japan.
3. Hancock, E. M.; Cope, A. C. *Org. Synth., Coll. Vol. III* **1965**, 219.
4. Butler, G. B.; Herring, K. H.; Lewis, P. L.; Sharpe, III, V. V.; Veazey, R. L. *J. Org. Chem.* **1977**, *42*, 679; Baucom, K. B.; Butler, G. B. *J. Org. Chem.* **1972**, *37*, 1730.
5. Boger, D. L.; Brotherton, C. E.; Georg, G. I. *Org. Synth.* **1987**, *65*, 32; Breslow, R.; Pecoraro, J.; Sugimoto, T. *Org. Synth., Coll. Vol. VI* **1988**, 361.
6. Review: Nakamura, E. In "Organic Synthesis in Japan. Past, Present and Future"; Noyori, R., Ed., Tokyo Kagaku Dojin, 1992; Nakamura, E.; Yamago, S *Acc. Chem. Res.* **2002**, *35*, 867; Yamago, S.; Nakamura, E. *Org. React.* **2002**, *61*, 1.
7. (a) Isaka, M.; Ejiri, S.; Nakamura, E. *Tetrahedron* **1992**, *48*, 2045; (b) Isaka, M.; Ando, R.; Morinaka, Y.; Nakamura, E. *Tetrahedron Lett.* **1991**, *32*, 1339; (c) Isaka, M.; Matsuzawa, S.; Yamago, S.; Ejiri, S.; Miyachi, Y.; Nakamura, E. *J. Org. Chem.* **1989**, *54*, 4727.

8. Review: Potts, K. T.; Baum, J. S. *Chem. Rev.* **1974**, *74*, 189; Eicher, T.; Weber J. L. *Top. Curr. Chem.* **1975**, *57*, 1.

9. Review: Nakamura, E. *J. Synth. Org. Chem., Jpn.* **1994**, *52*, 935-945; Nakamura, M.; Isobe, H.; Nakamura, E. *Chem. Rev.* submitted.

10. Tokuyama, H.; Isaka, M.; Nakamura, E.; Ando, R.; Morinaka, Y. *J. Antibiotics* **1992**, *45*, 1148.

11. Okuda, T.; Shimma, N.; Furumai, T. *J. Antibiotics* **1984**, *37*, 723; Okuda, T.; Furumai, T. *J. Antibiotics* **1985**, *38*, 631.

12. (a) Ando, R.; Morinaka, Y.; Tokuyama, H.; Isaka, M.; Nakamura, E. *J. Am. Chem. Soc.* **1993**, *115*, 1174; (b) Tokuyama, H.; Isaka, M.; Nakamura, E. *Synth. Commun.* **1995**, *25*, 2005.

13. Boger, D. L.; Brotherton-Pleiss, C. E. In "Advances in Cycloaddition"; JAI Press: Greenwich, **1990**; Vol. II, p. 147; Boger, D. L.; Brotherton, C. E. *J. Am. Chem. Soc.* **1986**, *108*, 6695.

14. Tokuyama, H.; Isaka, M.; Nakamura, E. *J. Am. Chem. Soc.* **1992**, *114*, 5523; Tokuyama, H.; Yamada, T.; Nakamura, E. *Synlett* **1993**, 589.

15. Isaka, M.; Nakamura, E. *J. Am. Chem. Soc.* **1990**, *112*, 7428; Nakamura, E.; Isaka, M.; Matsuzawa, S. *J. Am. Chem. Soc.* **1988**, *110*, 1297; Nakamura, M.; Arai, M.; Nakamura, E. *J. Am. Chem. Soc.* **1995**, *117*, 1179.

16. Albert, R. M.; Butler, G. B. *J. Org. Chem.* **1977**, *42*, 674.

17. Nakamura, E.; Isobe, H.; Tokuyama, H.; Sawamura, M. *J. Chem. Soc., Chem. Commun.* **1996**, 1747; Isobe, H.; Tokuyama, H.; Sawamura, M.; Nakamura, E. *J. Org. Chem.* **1997**, *62*, 5034; Isobe, H.; Tomita, N.; Jinno, S.; Okayama, H.; Nakamura, E. *Chem. Lett.* **2001**, 1214.

18. Yamago, S.; Nakamura, E. *J. Am. Chem. Soc.* **1989**, *111*, 7285.

19. Nakamura, E.; Yamago, S.; Ejiri, S.; Dorigo, A. E.; Morokuma, K. *J. Am. Chem. Soc.* **1991**, *113*, 3183.

20. Ejiri, S.; Yamago, S.; Nakamura, E. *J. Am. Chem. Soc.* **1992**, *114*, 8707.

21. Yamago, S.; Ejiri, S.; Nakamura, M.; Nakamura, E. *J. Am. Chem. Soc.* **1993**, *115*, 5344.

22. Prato, M.; Suzuki, T.; Foroudian, H.; Li, Q.; Khemani, K.; Wudl, F.; Leonetti, J.; Little, R. D.; White, T.; Rickbom, B.; Yamago, S.; Nakamura, E. *J. Am. Chem. Soc.* **1993**, *115*, 1594.

23. Yamago, S.; Nakamura, E. *Chem. Lett.* **1996**, 395.

24. Yamago, S.; Tokuyama, H.; Nakamura, E.; Kikuchi, K.; Kananishi, S.; Sueki, K.; Nakahara, H.; Enomoto, S.; Ambe, F. *Chemistry & Biology*, **1995**, *2*, 385.

25. Yamago, S.; Ejiri, S.; Nakamura, E. *Angew. Chem., Int. Ed. Engl.* **1995**, *34*, 2154.

26. Yamago, S.; Nakamura, E. *J. Org. Chem.* **1990**, *55*, 5553.

27. Yamago, S.; Nakamura, M.; Wang, X. Q.; Yanagawa, M.; Tokumitsu, S.; Nakamura, E. *J. Org. Chem.* **1998**, *63*, 1694.

28. Yamago, S.; Takeichi, A.; Nakamura, E. *J. Am. Chem. Soc.* **1994**, *116*, 1123; Yamago, S.; Takeichi, A.; Nakamura, E. *Synthesis* **1996**, 1380.

Appendix

Chemical Abstracts Nomenclature (Collective Index Number); (Registry Number)

2,2-Bis(chloromethyl)-5,5-dimethyl-1,3-dioxane: 1,3-Dioxane, 2,2-bis(chloromethyl)-5,5-dimethyl- (9); (133961-12-3]

2-Cyclopenten-1-one (8, 9); (930-30-3]

1,3-Dichloroacetone: 2-Propanone, 1,3-dichloro- (8, 9); (534-07-6)

6,6-Dimethyl-1-methylene-4,8-dioxaspiro[2.5]octane: 4,8-Dioxaspiro[2.5]octane, 6,6-dimethyl-1-methylene (9); (122968-05-2)

Ferric nitrate nonahydrate: Nitric acid, iron(3+) salt, nonahydrate (9); (7782-61-8)

Neopentyl glycol: 1,3-Propanediol, 2,2-dimethyl- (8, 9); (126-30-7)

Potassium tert-butoxide: 2-Propanol, 2-methyl-, potassium salt (9); (865-47-4)

Ammonia (8,9); (7664-41-7)

Sodium (8, 9); (7440-23-5)

Sodium amide: Sodium amide [$Na(NH_2)$] (9); (7782-92-5)

p-Toluenesulfonic acid: Benzenesulfonic acid, 4-methyl- (9); (104-15-4)

1,6,6-Trimethyl-4,8-dioxaspiro[2.5]oct-1-ene: 4,8-Dioxaspiro[2.5]oct-1-ene, 1,6,6-trimethyl- (9); (122762-81-6)

Methyl iodide: Methane, iodo- (8, 9); (74-88-4)

Table 1. Cycloaddition of **2** with 2π Electron Acceptors.[a]

Entry	Substrate	Product	Yield[b] (%)	Entry	Substrate	Product	Yield[b] (%)
1	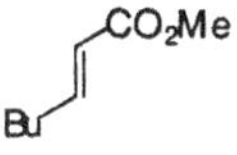 CO_2Me, Bu	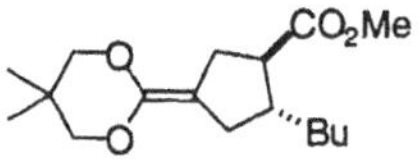 O, O, CO_2Me, Bu	89	**2**	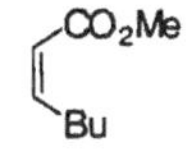 CO_2Me, Bu	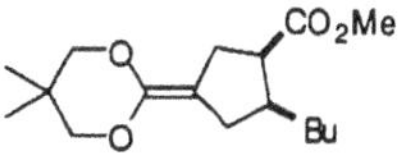 O, O, CO_2Me, Bu	86[c]
3	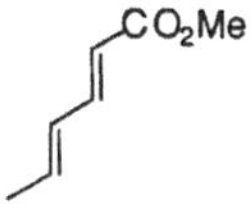 CO_2Me	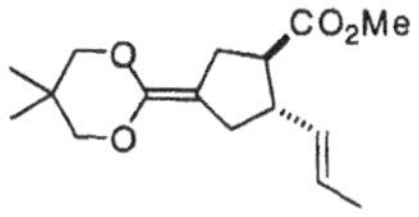 O, O, CO_2Me	90	**4**	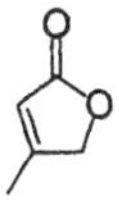 O, O	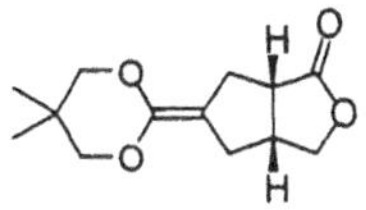 O, O, H, O, O, H	84
5	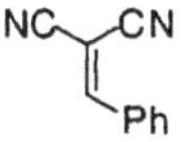 NC, CN, Ph	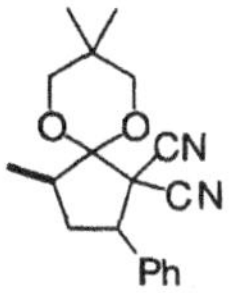 O, O, CN, CN, Ph	77	**6**	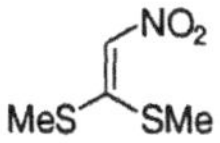 NO_2, MeS, SMe	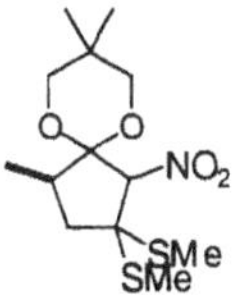 O, O, NO_2, SMe, SMe	42
7	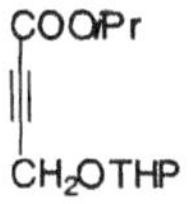 COOiPr, CH_2OTHP	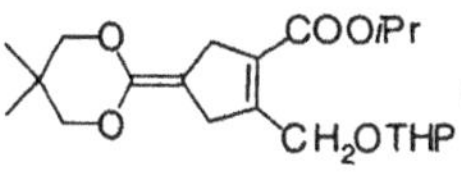 O, O, COOiPr, CH_2OTHP	61	**8**	 SOMe, $SiMe_3$	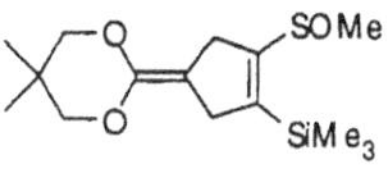 O, O, SOMe, $SiMe_3$	88
9	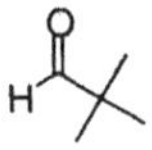 O, H	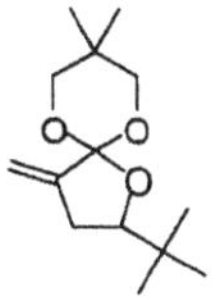 O, O, O	86	**10**	$(CH_2O)_n$[d]	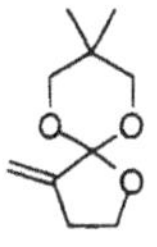 O, O, O	62
11	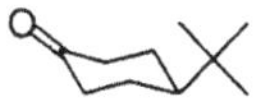 O	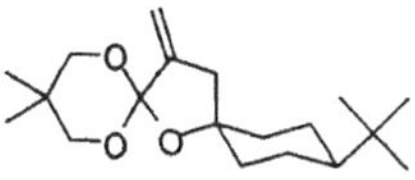 O, O, O	81[e]	**12**	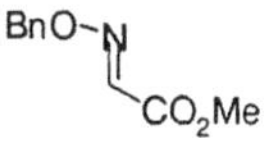 BnO–N, CO_2Me	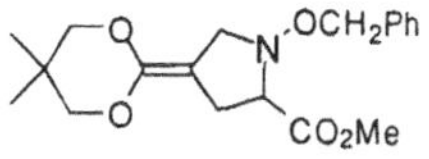 O, O, N, OCH_2Ph, CO_2Me	89

[a]The reactions were carried out at 40-130°C. [b]Isolated yield. The initially formed ketene acetal products were isolated as cyclopentane carboxylate esters after acid hydrolysis. [c]The starting olefin was 100% Z and the product was 98.3% Z. [d]$ZnCl_2$ (20 mol%) was added to the reaction mixture. [e]The product consisted of a 82:18 mixture of two diastereomers. The major isomer is shown.

GENERATION AND [2+2] CYCLOADDITIONS OF THIO-SUBSTITUTED KETENES:

trans-1-(4-METHOXYPHENYL)-4-PHENYL-3-(PHENYLTHIO)AZETIDIN-2-ONE

(2-Azetidinone, 1-(4-methoxyphenyl)-4-phenyl-3-(phenylthio)-, trans-)

A. PhCHO + 4-methoxyaniline (OCH_3, NH_2) →($MgSO_4$, CH_2Cl_2)→ PhCH=N–C$_6$H$_4$–OCH_3

B. PhS–C(=O)–CH_3 →((a) LiHMDS, THF, $CF_3CO_2CH_2CF_3$; (b) $C_{12}H_{25}$–C$_6$H$_4$–SO_2N_3, Et_3N, H_2O, CH_3CN)→ PhS–C(=O)–CH=N_2

C. PhCH=N–C$_6$H$_4$–OCH_3 + PhS–C(=O)–CH=N_2 →(cat. $Rh_2(OAc)_4$, CH_2Cl_2)→ trans-azetidin-2-one (PhS, Ph, N–C$_6$H$_4$–OCH_3)

Submitted by Rick L. Danheiser, Iwao Okamoto, Michael D. Lawlor, and Thomas W. Lee.[1]

Checked by Scott E. Denmark and Michael H. Ober.

1. Procedure

Caution! Diazo compounds are presumed to be toxic and potentially explosive and therefore should be handled with caution in a fume hood. Although in carrying out this reaction numerous times we have never observed an explosion, we recommend that these reactions be conducted behind a safety shield.

A. *N-Benzylidene-p-anisidine.* A 100-mL, three-necked, round-bottomed flask is equipped with an argon inlet adapter, rubber septum, glass stopper, and a magnetic stirring bar (Note 1). The flask is charged with 45 mL of dichloromethane (CH_2Cl_2) (Note 2) and 3.00 mL (0.030 mol) of benzaldehyde (Note 3), and then is cooled in an ice-water bath while a solution of 3.50 g (0.028 mol) of p-anisidine (Note 4) in 5 mL of CH_2Cl_2 is added dropwise via syringe over 15 min. After 30 min, 7.5 g of anhydrous magnesium sulfate is added in one portion. The ice-water bath is removed, and the reaction mixture is stirred at room temperature for 2 hr. The resulting mixture is then filtered through a sintered glass funnel with the aid of two 5-mL portions of CH_2Cl_2, and the filtrate is concentrated at reduced pressure by rotary evaporation at room temperature to afford a pale brown powder. This material is dissolved in 150 mL of ethanol heated in an 80°C water bath while 270 mL of hot water is added with stirring. The resulting solution is allowed to cool to room temperature and then is cooled in an ice-water bath for 2 hr. Filtration provides 5.31 g (88%) of N-benzylidene-p-anisidine as brown metallic plates (Note 5).

B. *S-Phenyl diazothioacetate.* A 250-mL, three-necked, round-bottomed flask is equipped with a magnetic stirring bar, argon inlet adapter, rubber septum, and a 50-mL pressure-equalizing dropping funnel fitted with a rubber septum (Note 1). The flask is charged with 50 mL of dry tetrahydrofuran (THF) (Note 6) and 12.0 mL (0.057 mol) of 1,1,1,3,3,3-hexamethyldisilazane (Note 7), and then is cooled in an ice-water bath while 20.7 mL (0.052 mol) of a 2.58M solution of n-butyllithium in hexane (Note 8) is added dropwise over 5 min by syringe. After 10 min, the resulting solution is cooled at -78°C in a dry ice-acetone bath, and a solution of 7.00 mL (0.052 mol) of S-phenyl thioacetate (Note 9) in 40 mL of dry tetrahydrofuran is added dropwise via the addition funnel over 30 min (the funnel is rinsed with two 2-mL portions of dry THF). The reaction mixture is allowed to stir for 30 min at -78°C, and then 8.50 mL (0.063 mol) of 2,2,2-trifluoroethyl trifluoroacetate

(Note 10) is added rapidly in one portion via syringe. After 10 min, the reaction mixture is poured into 200 mL of 5% aqueous hydrochloric acid and is extracted with three portions (250, 100, and 100 mL) of ether. The combined organic phases are washed with 100 mL of saturated sodium chloride solution, dried over anhydrous sodium sulfate, filtered, and concentrated under reduced pressure with a rotary evaporator, then under high vacuum to afford a colorless solid. This material is immediately dissolved in 100 mL of acetonitrile (Note 11) and is transferred into a 500-mL, three-necked, round-bottomed flask equipped with a magnetic stirring bar, argon inlet adapter, rubber septum, and 125-mL pressure-equalizing dropping funnel fitted with a rubber septum. Water (0.95 mL, 0.053 mol) and 10.0 mL (0.072 mol) of triethylamine (Note 12) are added via syringe, and a solution of 26.9 g (0.076 mol) of dodecylbenzenesulfonyl azide[2] (Note 13) in 45 mL of acetonitrile is then added through the dropping funnel over 15 min. After stirring at room temperature for 16 hr, the solution is concentrated at reduced pressure with a rotary evaporator and then a vacuum pump to give a brown oil. Silica gel (20 g) is added, and the resulting slurry is loaded onto a column (60 mm) of 200 g of silica gel (230-400 mesh) and is eluted with 5% ethyl acetate-hexane to afford 7.16 g (77%) of S-phenyl diazothioacetate as a pale yellow oil (Note 14).

C. *trans-1-(4-Methoxyphenyl)-4-phenyl-3-(phenylthio)azetidin-2-one.* A 500-mL, three-necked, round-bottomed flask is equipped with a magnetic stirring bar, reflux condenser fitted with an argon inlet adapter, glass stopper, and 50-mL, pressure-equalizing dropping funnel fitted with a glass stopper (Note 1). The flask is charged with 3.50 g (0.016 mol) of N-benzylidene-p-anisidine, 200 mL of CH_2Cl_2 (Note 2), and 0.045 g (0.10 mmol) of rhodium(II) acetate dimer (Note 15), and the resulting green solution is heated at reflux while a solution of 4.43 g (0.025 mol) of S-phenyl diazothioacetate in 40 mL of CH_2Cl_2 is added via the dropping funnel over 1 hr (the dropping funnel is rinsed with 2 mL of CH_2Cl_2). The reaction mixture is further heated at reflux for 5 min, then allowed to cool to room temperature. After transfer to a 500-mL, round-bottomed flask, the mixture is concentrated by rotary evaporation at reduced pressure to provide a brown oil. This material is filtered through a column of 50 g of silica gel (elution with 750 mL of CH_2Cl_2) to remove polar impurities and is concentrated under reduced pressure to afford a brown solid, which is washed on a sintered glass funnel with 10 mL of ethyl acetate and then 20 mL of hexane. The resulting pale yellow powder is dissolved in 40 mL of ethyl acetate at 80°C, and 400 mL of hexane (pre-heated in a water bath at 80°C) is then added in one portion. The resulting solution is allowed to cool to room temperature and then is cooled further at -20°C for 2 hr to afford 5.50 g (91%) of trans-1-(4-methoxyphenyl)-4-phenyl-3-(phenylthio)azetidin-2-one as off-white crystals (Notes 16 and 17).

2. Notes

1. The apparatus is flame-dried under reduced pressure and then maintained under an atmosphere of argon during the course of the reaction.

2. Dichloromethane was distilled from calcium hydride immediately before use.

3. Benzaldehyde was purchased from Aldrich Chemical Company, Inc., and was distilled before use.

4. p-Anisidine was purchased from Aldrich Chemical Company, Inc. and used as received.

5. The imine has the following physical properties: mp 70-71°C (lit.[3] 70-71°C); ^{1}H NMR (500 MHz, $CDCl_3$) δ: 3.83 (s, 3 H), 6.93 (d, 2 H, J = 9.0), 7.24 (d, 2 H, J = 8.8), 7.46 (m, 3 H), 8.48 (s, 1 H), 7.89 (m, 2 H); ^{13}C NMR (166 MHz, $CDCl_3$) δ: 55.7, 114.6, 122.4, 128.8, 128.9, 131.3, 136.6, 145.2, 158.5, 158.7; IR (CCl_4) cm^{-1}: 3065, 3030, 3002, 2959, 2935, 2909, 2880, 2835, 1626, 1529, 1505, 1464, 1366, 1289, 1245, 1191, 1106, 1040, 920, 882, 828, 761, 691, 542; Anal. Calcd for $C_{14}H_{13}NO$: C, 79.59; H, 6.20; N, 6.63. Found: C, 79.50; H, 6.05; N, 6.74.

6. Tetrahydrofuran was distilled from sodium benzophenone ketyl immediately before use.

7. 1,1,1,3,3,3-Hexamethyldisilazane was purchased from Aldrich Chemical Company, Inc., and was distilled from calcium hydride prior to use.

8. n-Butyllithium was purchased from Fisher Scientific Company, and was titrated prior to use according to the method of Watson and Gilman.[4]

9. S-Phenyl thioacetate was purchased from Aldrich Chemical Company, Inc. and used as received.

10. 2,2,2-Trifluoroethyl trifluoroacetate was purchased from Aldrich Chemical Company, Inc., and was distilled from calcium hydride prior to use.

11. Acetonitrile was distilled from calcium hydride prior to use.

12. Triethylamine was purchased from EM Science, and was distilled from calcium hydride prior to use.

13. Dodecylbenzenesulfonyl azide was purchased by the checkers from TCI Americas Inc. and was used as received.

14. S-Phenyl diazothioacetate decomposes slowly on storage at -20°C with the generation of nitrogen gas. The diazo compound is best used immediately in the next step or purified by chromatography immediately before use. Physical properties are as follows: ^{1}H NMR (500 MHz, $CDCl_3$) δ: 5.25 (s, 1 H), 7.40-7.45 (m, 3 H), 7.49-7.51 (m, 2 H); ^{13}C NMR (166 MHz, $CDCl_3$) δ: 54.5, 127.6, 129.6, 130.1, 135.6, 184.3; IR (thin film) cm^{-1}: 3099, 3077, 2466, 2354, 2269, 2108, 1641, 1477, 1441, 1334, 1139, 1022, 854, 748, 690, 638.

15. Rhodium(II) acetate dimer was purchased from Alfa Aesar Company and used as received.

16. Physical properties of the β-lactam are as follows: mp 144-146°C (lit. 142°C[5], 147-148°C[6]); ^{1}H NMR (500 MHz, $CDCl_3$) δ: 3.72 (s, 3 H), 4.25 (d, 1 H, J = 2.4), 4.77 (d, 1 H, J = 2.4), 7.52-7.51 (m, 2 H), 7.36-7.34 (m, 3 H), 7.31-7.26 (m, 5 H), 7.16 (d, 2 H,

J = 9.2), 6.75 (d, 2 H, J = 9.2),; ^{13}C NMR (166 MHz, $CDCl_3$) δ: 162.9, 156.5, 136.6, 132.5, 130.9, 129.6, 129.5, 129.4, 129.2, 128.2, 126.3, 118.8, 114.5, 63.3, 61.7, 55.6; IR (CCl_4) cm^{-1}: 3065, 3033, 3003, 2953, 2934, 2909, 2835, 1760, 1512, 1455, 1381, 1248, 1179, 1137, 1067, 1040, 828, 736, 696, 590; Anal. Calcd for $C_{22}H_{19}NO_2S$: C, 73.10; H, 5.30; N, 3.88; S, 8.87. Found: C, 73.03; H, 5.29; N, 4.02; S, 9.20.

17. Cleavage of the p-methoxyphenyl group can be achieved in 65% yield by the following procedure. A three-necked, 250-mL, round-bottom flask, equipped with a magnetic stirbar, rubber septum, argon inlet adapter, and glass stopper, is charged with 0.512 g (1.42 mmol) of trans-1-(4-methoxyphenyl)-3-(phenylthio)azetidin-2-one and 55 mL of acetonitrile. After cooling the flask in an ice-water bath, a solution of 2.380 g (4.34 mmol) of ceric ammonium nitrate in 20 mL of water is added by cannula over 5 min followed by stirring for 1 hr. The reaction mixture is poured into 500 mL of water and is extracted with three 200-mL portions of ethyl acetate. The organic phases are combined and washed with 400 mL of 5% aqueous $NaHCO_3$ solution. The washes are back extracted with 100 mL of ethyl acetate. The combined organic phases are washed additionally with six 200-mL portions of saturated aqueous Na_2SO_3 solution, then two 150-mL portions of saturated aqueous NaCl solution. After drying with Na_2SO_4, the solution is concentrated to give a brown oil. Silica gel (1.5 g) is added, and the resulting slurry is applied to a column of 10 g of silica gel and eluted with 10% ethyl acetate-hexane. The eluent is concentrated to provide a yellow solid which is dissolved in 125 mL of boiling hexane, which is then reduced to a total volume of 50 mL. The resulting solution is allowed to cool to room temperature, then is cooled further to -20°C to afford 0.234 g (65%) of 4-phenyl-3-(phenylthio)-2-azetidinone as a pale yellow solid, mp 111-112°C (lit.[7] 110.5-111.5°C).

Waste Disposal Information

All toxic materials were disposed of in accordance with "Prudent Practices in the Laboratory"; National Academy Press; Washington, DC, 1995.

3. Discussion

[2+2] Cycloadditions of ketenes with alkenes and alkynes constitute the most popular method for the synthesis of cyclobutanones and cyclobutenones. Unfortunately, however, this process is truly general only for highly nucleophilic ketenophiles such as conjugated dienes and enol ethers. In general, unactivated alkenes and alkynes fail to react in good yield with either alkyl- or aryl-substituted ketenes, or with ketene itself. To circumvent this limitation, dichloroketene is usually employed as a ketene equivalent, since this electrophilic ketene reacts well with many types of unactivated multiple bonds, and the resultant cycloadducts undergo facile dechlorination under mild conditions.[8]

The procedure described here illustrates a practical and convenient method for the generation of thio-substituted ketenes which participate in surprisingly facile cycloadditions with both activated and unactivated alkenes, alkynes, and imines to form four-membered carbocycles and heterocycles. As outlined in the following scheme, exposure of α-diazo thiol esters to the action of catalytic rhodium(II) acetate leads to a remarkably facile "thia-Wolff rearrangement," producing thio-substituted ketenes which combine with a variety of ketenophiles to provide access to α-thiocyclobutanones, cyclobutenones, and β-lactams.

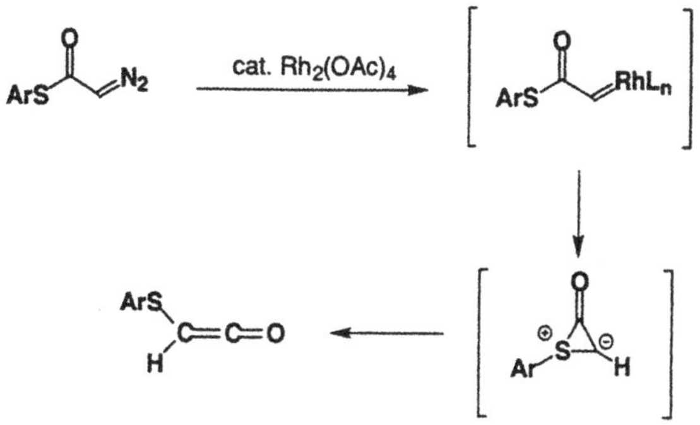

As illustrated below, desulfurization of α-thiocyclobutanone cycloadducts is easily achieved in high yield upon exposure to either tributyltin hydride or activated zinc dust, and the overall sequence thus represents a useful new alternative to the existing dichloroketene-based methodology for the synthesis of a variety of four-membered carbocycles and heterocycles[9].

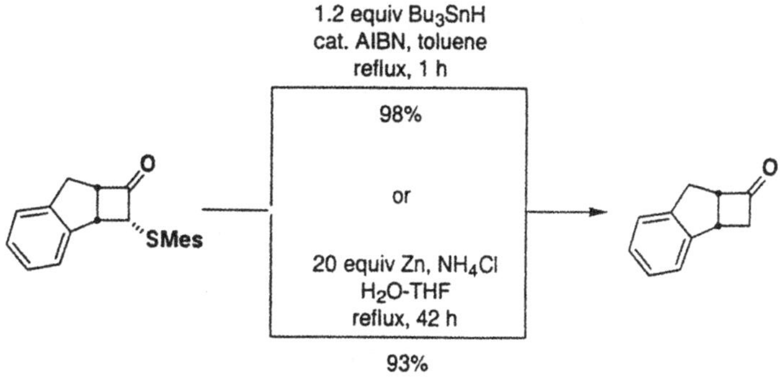

The requisite α-diazo thiol esters are conveniently prepared by using the "detrifluoroacetylative" diazo transfer strategy previously developed in our laboratory.[10] Cycloadditions are best carried out by using as little as 0.006 equiv of rhodium(II) acetate to promote the thia-Wolff rearrangement. Reactions involving the more nucleophilic ketenophiles proceed smoothly in refluxing dichloromethane (40°C), while cycloadditions with less reactive partners are best accomplished in 1,2-dichloroethane (83°C). As is standard for ketene cycloadditions, the optimal protocol involves slowly adding a solution of the diazo thiol ester to a solution of the ketenophile and catalyst in order to minimize competitive ketene dimerization.

Examples of the application of this chemistry to the preparation of cyclobutanones, cyclobutenones, and β-lactams are presented in the Table. The mesityl thiol ester has proven to be particularly effective in reactions with less ketenophilic alkenes, although with the more reactive ketenophiles nearly identical results are obtained using either the mesityl α-diazo thiol ester or the more readily available thiophenyl ester. In the case of readily available ketenophiles, the reaction is best conducted using excess alkene, alkyne, or imine, but in other cases the cycloaddition can be carried out with excess diazo thiol ester. The efficiency of the reaction with unactivated alkenes is especially notable, and compares favorably with results obtained previously employing dichloroketene. For example, addition of dichloroketene to methylenecyclohexane is reported to proceed in 55% yield,[11] while up to 81% of the desired [2+2] cycloadduct is produced in the reaction of (mesitylthio)ketene with this olefin under our conditions.

1. Department of Chemistry, Massachusetts Institute of Technology, Cambridge, MA 02139. We thank the National Institutes of Health (GM 28273) for generous financial support. M.D.L. was supported in part by NIH training grant CA 09112 and I. O. was supported as a J.S.P.S. Research Fellow. T.W.L. was a James Flack Norris Summer Undergraduate Research Scholar of the Northeastern Section of the American Chemical Society and a Pfizer Undergraduate Summer Research Fellow.

2. Hazen, G. G.; Bollinger, F. W.; Roberts, F. E.; Russ, W. K.; Seman, J. J.; Staskiewicz, S. *Org. Synth. Coll. Vol. IX* **1998**, 400.
3. Roe, A.; Montgomery, J. A. *J. Am. Chem. Soc.* **1953**, *75*, 910.

4. Gilman, H.; Cartledge, F. K. *J. Organomet. Chem.* **1964**, *2*, 447.
5. van der Veen, J. M.; Bari, S. S.; Krishnan, L.; Manhas, M. S.; Bose, A. K. *J. Org. Chem.* **1989**, *54*, 5758.
6. Cossío, F. P.; Ganboa, I.; García, J. M.; Lecea, B.; Palomo, C. *Tetrahedron Lett.* **1987**, *28*, 1945.
7. Hart, D. J.; Kanai, K.; Thomas, D. G.; Yang, T. -K. *J. Org. Chem.* **1983**, *48*, 289.
8. For recent reviews, see: (a) Hyatt, J. A.; Raynolds, P. W. *Org. Reactions* **1994**, *45*, 159. (b) Tidwell, T. T. *Ketenes*; Wiley: New York, 1995. (c) Schaumann, E.; Scheiblich, S. In *Methoden der Organischen Chemie (Houben-Weyl)*; Kropf, H., Schaumann, E., Eds.; Thieme: Stuttgart, Germany, 1993; Vol. E15, Parts 2 and 3. (d) Lee-Ruff, E. In *Methods of Organic Chemistry (Houben-Weyl)*; de Meijere, A., Ed.; Thieme: Stuttgart, Germany, 1997; Vol. E17e, pp 190-219.
9. Lawlor, M. D.; Lee, T. W.; Danheiser, R. L. *J. Org. Chem.* **2000**, *65*, 4375.
10. (a) Danheiser, R. L.; Miller, R. F.; Brisbois, R. G.; Park, S. Z. *J. Org. Chem.* **1990**, *55*, 1959. (b) Danheiser, R. L.; Miller, R. F.; Brisbois, R. G. *Org. Synth. Coll. Vol. IX* **1998**, 197.
11. Dunkelblum, E. *Tetrahedron* **1976**, *32*, 975.

TABLE
Preparation of Cyclobutanones and β-Lactams from α-Diazo Thiol Esters

ketenophile	diazo thiol ester	reaction conditions[a]	product	% yield (ratio)[b]
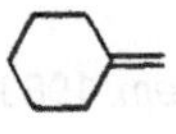	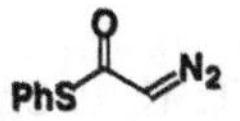 O, PhS, N_2	A	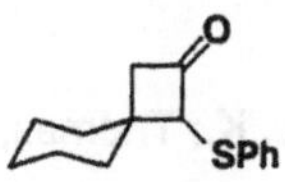 O, SPh	73-78
	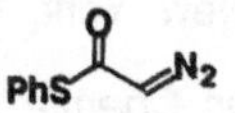 O, PhS, N_2	B	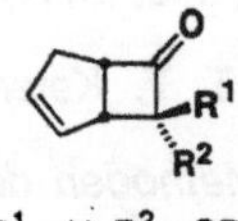 O, R^1, R^2 R^1 = H, R^2 = SPh R^1 = SPh, R^2 = H	90-96 (5-11:1)
 Ph	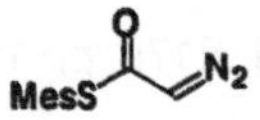 O, MesS, N_2	A	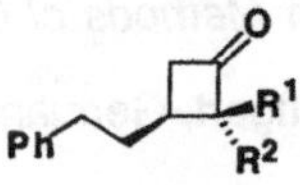 O, Ph, R^1, R^2 R^1 = H, R^2 = SMes R^1 = SMes, R^2 = H	62 (83:17)
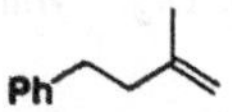 Ph	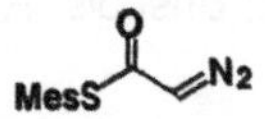 O, MesS, N_2	A	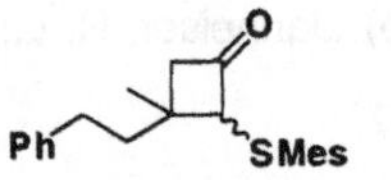 O, Ph, SMes	59 (57:43)
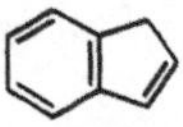	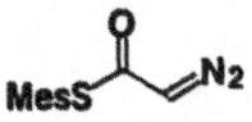 O, MesS, N_2	A	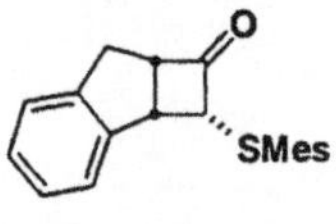 O, SMes	58
PhCH=NPh	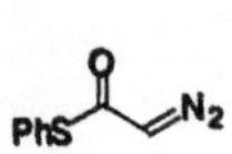 O, PhS, N_2	B	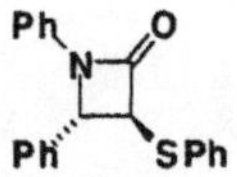 Ph, N, O, Ph, SPh	85
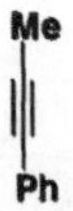 Me, Ph	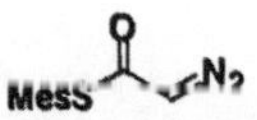 O, MesS, N_2	A	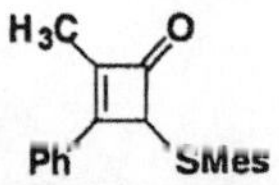 H_3C, O, Ph, SMes	50

[a] The diazo thiol ester is treated with 0.02 equiv of $Rh_2(OAc)_4$ and 3.5-10 equiv of ketenophile. Procedure A: reaction in 1,2-dichloroethane at 83 °C for 3 h. Procedure B: reaction in dichloromethane at 40 °C for 3 h. [b] Isolated yield and ratio of diastereoisomers.

Appendix

Chemical Abstracts Nomenclature (Collective Index Number); (Registry Number)

trans-1-(4-Methoxyphenyl)-4-phenyl-3-(phenylthio)azetidin-2-one: 2-Azetidinone, 1-(4-methoxyphenyl)-4-phenyl-3-(phenylthio)-, trans- (9); (94612-48-3)

Benzaldehyde (8,9); (100-52-7)

p-Anisidine (8); Benzenamine, 4-methoxy- (9); (104-94-9)

N-Benzylidene-p-anisidine: Benzenamine, 4-methoxy-N-(phenylmethylene)- (9); (783-08-4)

1,1,1,3,3,3-Hexamethyldisilazane: Silanamine, 1,1,1-trimethyl-N-(trimethylsilyl)- (9); (999-97-3)

Butyllithium: Lithium, butyl- (8,9); (109-72-8)

S-Phenylthioacetate: Ethanethioic acid, S-phenyl ester (9); (934-87-2)

2,2,2-Trifluoroethyl trifluoroacetate: Acetic acid, trifluoro-, 2,2,2-trifluoroethyl ester (8,9); (407-38-5)

Triethylamine (8); Ethanamine, N,N-diethyl- (9); (121-44-8)

4-Dodecylbenzenesulfonyl azide: Benzenesulfonyl azide, 4-dodecyl- (13); (79791-38-1)

S-Phenyldiazothioacetate: Ethanethioic acid, diazo-S-phenyl ester (13); (72228-26-3)

Rhodium(II) acetate dimer: Acetic acid, rhodium(2+) salt (8,9); (5503-41-3)

PREPARATION AND USE OF A NEW DIFLUOROCARBENE REAGENT. TRIMETHYLSILYL 2-FLUOROSULFONYL-2,2-DIFLUOROACETATE: n-BUTYL 2,2-DIFLUOROCYCLOPROPANECARBOXYLATE

(Cyclopropanecarboxylic acid, 2,2-difluoro-, butyl ester and Acetic acid, difluoro(fluorosulfonyl)-, trimethylsilyl ester)

A. $FSO_2CF_2CO_2H + (CH_3)_3SiCl \xrightarrow{0° \text{ to RT}} FSO_2CF_2CO_2Si(CH_3)_3$

B. $CH_2{=}CHCO_2Bu + FSO_2CF_2CO_2Si(CH_3)_3 \xrightarrow[\text{toluene, reflux}]{\text{NaF (cat)}}$ 2,2-difluoro-1-(CO_2Bu)-cyclopropane (F, F, CO_2Bu, H)

Submitted by W. R. Dolbier, Jr.,[1] F. Tian, J.-X. Duan, and Q.-Y. Chen.

Checked by Victoria D. Bock, David P. Richardson and Steven Wolff.

1. Procedure

Caution: Trimethylsilyl 2-fluorosulfonyl-2,2-difluoroacetate should be handled with care. Skin contact, which will cause painful burns and blistering, should be avoided. If such contact occurs, the affected area should be washed thoroughly with water and with sodium bicarbonate solution.

A. *Trimethylsilyl 2-fluorosulfonyl-2,2-difluoroacetate.* All glassware is oven-dried for about 1 hr prior to use. A 1-L, three-necked, round-bottomed flask is equipped with a magnetic stirrer, addition funnel with a nitrogen (N_2) inlet, and water-cooled condenser with gas outlet. The gas outlet is connected by Tygon tubing to an empty 500-mL back-up trap, and then to an inverted 75-mm glass funnel outlet positioned just above a 1-L beaker containing 75 g of sodium bicarbonate ($NaHCO_3$) in 500 mL water (to neutralize HCl and chlorotrimethylsilane from the reaction vessel) (Note 1). The flask is charged with 150 g of 2-fluorosulfonyl-2,2-difluoroacetic acid (0.84 mol) (Note 2). Then, with a slow N_2 flow and cooling with an ice bath, 363 g of chlorotrimethylsilane (3.2 mol) (Notes 3, 4) is added dropwise with stirring over a 2 hr period. Upon completion of the addition, the mixture is allowed to warm to room temperature and is stirred with N_2 bubbling for 24 hr. The product mixture is distilled to give 175 grams (83%) of trimethylsilyl 2-fluorosulfonyl-2,2-difluoroacetate (bp 52°C at 15 mm) (Notes 5, 6).

B. *n-Butyl 2,2-difluorocyclopropanecarboxylate.* A 500-mL, round-bottomed, three-necked flask is fitted with a magnetic stirrer, condenser, addition funnel, gas dispersion tube extending to the bottom of the flask, and gas outlet with a paraffin oil bubbler. The flask is charged with 200 mL of toluene, 0.4 g of sodium fluoride (0.06 eq), and 20 g of n-butyl acrylate (0.156 mol) (Notes 7, 8). The solution is heated to reflux and slow N_2 bubbling is initiated with stirring for 1 hr. Trimethylsilyl 2-fluorosulfonyl-2,2-difluoroacetate (62.5 g, 0.31 mol, 1.6 eq) (Note 5) is added dropwise (Note 9). The mixture is heated for 8 hr, then cooled and filtered under vacuum filtration through a Celite pad (Note 10). Toluene is removed by simple distillation at atmospheric pressure, and the residue distilled at reduced pressure to obtain 15.4 g of n-butyl 2,2-difluorocyclopropanecarboxylate (55%) (bp 99-101°C at 58 mm) as a colorless liquid (Note 11).

2. Notes

1. In order to avoid suction of the $NaHCO_3$ solution into the trap, the mouth of the funnel should not be submerged in the $NaHCO_3$ solution.

2. 2-Fluorosulfonyl-2,2-difluoroacetic acid can be obtained from Aldrich Chemical Co., Inc., Oakwood Products, Inc., or from FluoroTech, LLC (PO Box 13135, Gainesville, FL 32604).

3. A 3.5-4 fold excess of chlorotrimethylsilane is required for full conversion of the 2-fluorosulfonyl-2,2-difluoroacetic acid. Residual acid can be difficult to separate from the ester.

4. Chlorotrimethylsilane (bp 55°C) was obtained from Aldrich Chemical Co., Inc. and used as received.

5. Trimethylsilyl 2-fluorosulfonyl-2,2-difluoroacetate is moisture-sensitive and, if possible, should be prepared immediately before use in the next step.

6. Trimethylsilyl 2-fluorosulfonyl-2,2-difluoroacetate is characterized as follows: ^{1}H NMR ($CDCl_3$) δ: 0.40 ppm (s); ^{13}C NMR, δ: -1.05 ppm (s), 112.22 (dt, J = 31.5, 299.0), 155.13 (t, J = 27.0); ^{19}F NMR, δ: -103.74 ppm (2F, s), 40.58 (1F, s).[2]

7. Toluene (anhydrous), sodium fluoride (Reagent grade), and n-butyl acrylate (99+%) were obtained from Aldrich Chemical Co., Inc. and used as received.

8. Other acrylate esters should function equally well in this synthesis. The n-butyl ester was specifically chosen because the boiling point of the product is sufficiently different from that of toluene, permitting convenient separation by distillation.

9. Addition of the trimethylsilyl 2-fluorosulfonyl-2,2-difluoroacetate should be slow at the beginning of the reaction, but can be gradually increased. A vigorous evolution of gas (CO_2 and SO_2) accompanies the early stages of addition, but will be barely distinguishable from the nitrogen flow at the end.

10. Filtration through the Celite pad is recommended in order to remove small amounts of acrylate polymer.

11. n-Butyl 2,2-difluorocyclopropanecarboxylate is characterized as follows: ^{1}H NMR ($CDCl_3$) δ: 0.91 (3H, t, J = 7.32), 1.39 (2H, m), 1.62 (2H, m), 1.72 (1H, m), 2.04 (1H, m), 2.41 (1H,m), 4.15 (2H, q, J = 6.75); ^{19}F NMR, δ: -126.50 (1F, m), -114.70 (1F,m); ^{13}C NMR, δ: 13.49, 16.30 (t, J = 10.6), 18.92, 25.51 (t, J = 10.6), 30.43, 65.31, 110.6 (m), 166.57 ppm. The product obtained by the checkers was contaminated with small amounts of n-butyl acrylate, which could not be removed by distillation.

Waste Disposal Information

All waste materials were disposed of in accordance with "Prudent Practices in the Laboratory"; National Academy Press; Washington, DC, 1995.

3. Discussion

The present example provides a typical procedure for difluorocyclopropanation utilizing a new and highly versatile source of difluorocarbene.[3,4] This novel reagent, trimethylsilyl 2-fluorosulfonyl-2,2-difluoroacetate, provides a convenient procedure for the synthesis of difluorocyclopropane compounds via addition of difluorocarbene to alkenes of highly variable nucleophilicity.[2] A relatively small excess of reagent is required for high yields, and even electrophilic alkenes, such as n-butyl acrylate, provide acceptable yields of the difluorocyclopropane adduct. The reactivity and preparative utility of our reagent is probably comparable to Seyferth's reagent, phenyl(trifluoromethyl)mercury,[5] but this latter substance is no longer commercially available and is both tedious and hazardous to prepare.

1. Department of Chemistry, University of Florida, PO Box 117200, Gainesville, FL 32611-7200 (email: wrd@chem.ufl.edu).
2. Tian, F.; Kruger, V.; Bautista, O.; Duan, J.-X.; Li, A.-R.; Dolbier, W. R., Jr.; Chen, Q.-Y. *Org. Lett.* **2000**, *2*, 563.
3. Brahms, D. L. S.; Dailey, W. P. *Chem. Rev.* **1996**, *96*, 1585.
4. Burton, D. J.; Hahnfeld, J. L. *Fluorine Chem. Rev.* **1977**, *8*, 119.
5. Seyferth, D.; Hopper, S. P. *J. Org. Chem.* **1972**, *37*, 4070.

Appendix

Chemical Abstracts Nomenclature (Collective Index Number) (Registry Number)

n-Butyl 2,2-difluorocyclopropanecarboxylate: Cyclopropanecarboxylic acid, 2,2-difluoro-, butyl ester (9); (260352-79-2)

Trimethylsilyl 2-fluorosulfonyl-2,2-difluoroacetate: Acetic acid, difluoro(fluorosulfonyl)-, trimethylsilyl ester (9); (120801-75-4)

Chlorotrimethylsilane: Silane, chlorotrimethyl- (8,9); (75-77-4)

2-Fluorosulfonyl-2,2-difluoroacetic acid: Acetic acid, difluoro(fluorosulfonyl)- (8,9); (1717-59-5)

n-Butyl acrylate: 2-Propenoic acid, butyl ester (9); (141-32-2)

Sodium fluoride: Sodium fluoride (NaF) (9); (7681-49-4)

PREPARATION OF α-ACETOXY ETHERS BY THE REDUCTIVE ACETYLATION OF ESTERS: endo-1-BORNYLOXYETHYL ACETATE

(Ethanol, 1-[[(1R,2S,4R)-1,7,7-trimethylbicyclo[2.2.1]hept-2-yl]oxy]-, acetate)

1. DIBALH CH_2Cl_2, -78°C, 45 min
2. pyr, DMAP, Ac_2O CH_2Cl_2, -78°C to 0°C, 12 hr

Submitted by David J. Kopecky and Scott D. Rychnovsky.[1]

Checked by Guillaume L. N. Péron, Cédric R. N. Fischmeister and Andrew B. Holmes.

1. Procedure

A flame-dried, 2-L, three-necked, round-bottomed flask, equipped with a large Teflon magnetic stirring bar and three rubber septa, is charged with 5.5 g (28.1 mmol) of (−)-bornyl acetate (Note 1) in 162 mL of dichloromethane (Note 2) under an argon atmosphere. A thermocouple (Note 3) is immersed in the solution through one of the rubber septa. The flask is placed in a dry ice/acetone bath and cooled to an internal solution temperature of −78°C. Stirring is begun and diisobutylaluminum hydride, 54.4 mL of a 1.0M solution in hexanes (54.4 mmol, 2 eq), is added as a slow stream via a glass syringe over a period of 13 min; during this addition, the internal solution temperature does not exceed −72°C (Notes 1,4). The clear solution is stirred for an additional 45 min at −78°C, then 6.6 mL of pyridine (6.5 g, 81.6 mmol, 3.0 eq) is added dropwise via a glass syringe, maintaining the internal temperature below −76°C. Immediately following, a solution of 6.5 g of 4-dimethylaminopyridine (DMAP) (54.5 mmol, 2.0 eq) in 80 mL of dichloromethane (Note 2) is added slowly via an 18-gauge stainless steel cannula (Note 5). The internal temperature is not allowed to rise above −72°C during this addition. Acetic anhydride

(15.4 mL, 16.6 g, 163.2 mmol, 6.0 eq) is then added dropwise via a glass syringe over 5.5 min at a rate to maintain the internal temperature below –72°C.

The resultant light yellow solution is stirred at –78°C for 12 hr (Note 6). During this time, the color of the solution intensifies considerably. The bright yellow solution is warmed to 0°C in an ice water bath and stirred for 35 min, during which time the color of the solution becomes bright yellow-orange. The reaction is then quenched by the sequential addition of 270 mL of saturated aqueous ammonium chloride and 200 mL of saturated aqueous sodium potassium tartrate. The emulsion is removed from the ice water bath and stirred vigorously for 50 min, by which time adequate phase separation is observed (Note 7). The biphasic mixture is transferred to a 2-L separatory funnel and is extracted with four 175-mL portions of dichloromethane. The combined extracts are washed sequentially with two 500-mL portions of an ice-cooled 1M aqueous solution of sodium bisulfate, three 500-mL portions of saturated aqueous sodium bicarbonate and 500 mL of saturated aqueous sodium chloride. The solution is dried over anhydrous sodium sulfate, filtered, and is concentrated under reduced pressure to afford 8.5 g of a yellow oil.

The residue is purified using flash chromatography with 145 g of triethylamine-deactivated Merck 9385 silica gel 60 (230-400) (Note 8) eluting with 5% ethyl ether in hexanes (Note 9). Fractions of 50-mL are collected, and product-containing fractions are identified by TLC analysis (Note 10). The product (6.4 g, 95%) is obtained as a clear oil (Notes 11,12). ^{1}H NMR and ^{13}C NMR spectra indicate that the product is a 2.8:1 mixture of diastereomers. Short path vacuum distillation of a portion of the product afforded material of comparable purity, bp 73.5-75°C (2 mm).

2. Notes

1. (–)-Bornyl acetate, purchased from Aldrich Chemical Company, Inc., was of 97% purity and was used as received. Diisobutylaluminum hydride (DIBALH) (1.0M solution in hexanes) and 4-dimethylaminopyridine were purchased from Aldrich Chemical Company, Inc. and used as received. Pyridine was purchased from Aldrich Chemical Company, Inc. and was distilled from calcium hydride at atmospheric pressure prior to use. Acetic anhydride was purchased from Fisher Scientific, Inc. and was distilled at atmospheric pressure prior to use.

2. The checkers obtained dry dichloromethane by distillation from calcium hydride. The submitters dried dichloromethane through an alumina filtration system[2].

3. A Digi-Sense digital thermometer was used. The internal temperature needs to be monitored, and the additions should be slow enough to avoid significant exotherms. The checkers found that temperature control is the major factor influencing the outcome of the reaction. The initial procedure was submitted as double scale [*i.e.* 11 g (56.2 mmol) (–)-bornyl acetate]. However, the checkers found temperature control on full scale to be difficult, resulting in incomplete reaction and/or over-reduction of the bornyl acetate. The checkers found that the procedure is reliably reproducible on half scale (5.5 g (-)-bornyl acetate).

4. The checkers found that reproducible results were obtained when a freshly opened bottle of diisobutylaluminium hydride was used.

5. The cannula transfer was accomplished by briefly evacuating the reaction flask using a vacuum line connected to a vacuum pump, and then pressurizing the flask containing the DMAP solution with just over 1 atmosphere of argon.

6. A bath temperature of –78°C was maintained during the 12 hr period by packing the bath with dry ice and covering the bath and reaction flask with aluminum foil.

7. Although the two phases separated, the aqueous phase remained cloudy.

8. The submitters used ICN 230-400 mesh silica gel. The silica gel was deactivated by mixing it with 325 mL of 2% (v/v) triethylamine in hexanes. The resultant slurry was poured onto the column, packed, and then flushed with 200 mL of 5% ethyl ether in hexanes to remove residual triethylamine before product loading.

9. The head pressure of the flash column during chromatography was 2.5 psi.

10. The α-acetoxy ether elutes with an R_f of 0.32 on Merck silica gel F_{254} in 5% (v/v) ethyl ether in hexanes (product stains yellow-green in an acidic solution of p-anisaldehyde).

11. The submitters obtained the equivalent of 6.4 g (95%). Physical and spectroscopic properties are as follows: IR (mixture of isomers, neat) cm^{-1}: 2952, 2879, 1737, 1452, 1372, 1247, 1175, 1141, 1072, 1033, 1009, 929, 833; ^{1}H NMR (500 MHz, $CDCl_3$, mixture of isomers) δ: 0.81-0.85 (m, 9 H), 0.99 (dd, 1 H, J = 13.1, 3.4, minor isomer), 1.10 (dd, 1 H, J = 13.3, 3.4, major isomer), 1.16-1.28 (m, 2 H), 1.38 (d, 3 H, J = 5.4, minor), 1.40 (d, 3 H, J = 5.3, major), 1.59 (t, 1 H, J = 4.5, major). 1.64 (t, 1 H, J = 4.5, minor), 1.63-1.74 (m, 1 H), 1.90-2.01 (m, 1 H), 2.04 (s, 3 H, major), 2.05 (s, 3 H, minor), 2.07-2.19 (m, 1 H), 3.81 (ddd, 1 H, J = 9.9, 3.2, 2.0), 5.93 (q, 1 H, J = 5.2, minor), 5.94 (q, 1 H, J = 5.2, major); ^{13}C NMR (125 MHz, $CDCl_3$, mixture of isomers) δ: 13.4 (minor isomer), 13.6 (major isomer), 18.8 (major), 18.8 (minor), 19.7 (major), 19.7 (minor), 20.9 (major), 21.2 (minor), 21.4 (minor), 21.6 (major), 26.4 (minor), 26.6 (major), 28.1 (major), 28.2 (minor), 36.1 (minor), 37.0 (major), 44.9 (minor), 45.0 (major), 47.2 (major), 47.6 (minor), 48.8 (minor), 49.3 (major), 82.6 (minor), 85.9 (major), 95.3 (minor), 97.7 (major), 170.6 (major), 170.9 (minor). Anal. Calcd for $C_{14}H_{24}O_3$: C, 70.0; H, 10.1. Found: C, 70.0; H, 10.0. The submitters obtained C, 70.0; H, 10.0.

12. The product contained a 3% impurity of (-)-bornyl acetate as indicated by the presence of the following diagnostic peaks in the ^{1}H NMR spectrum: 0.87 (s, 3 H), 0.90 (s, 3 H), 0.96 (buried dd, 1 H), 2.06 (s, 3 H). This impurity originates from incomplete reduction and/or over-reduction.

Waste Disposal Information

All toxic materials were disposed of in accordance with "Prudent Practices in the Laboratory"; National Academy Press; Washington, DC, 1995.

3. Discussion

The reductive acetylation of esters and lactones can be accomplished by treatment of the ester with a two-fold excess of DIBALH at –78°C followed by trapping of the resulting aluminum hemiacetal intermediate with acetic anhydride in the presence of pyridine and DMAP at low temperature.[3] This transformation has been successfully applied to a wide range of esters and lactones, including sterically-hindered acyclic esters, β-alkoxy esters, benzyl esters and macrolactones. A compilation of representative examples is presented in Table I.[3a] Upon exposure to mild acid, the α-acetoxy ether products undergo facile conversion to the corresponding oxacarbenium ions, which can be trapped by a wide variety of carbon- or heteroatom-based nucleophiles.[3b] For example, (endo)-1-bornyloxyethyl acetate can be efficiently allylated by treatment with boron trifluoride etherate and allyltrimethylsilane (82% yield; d.r. = 3.7:1), or ethylated with trimethylsilyl triflate and diethylzinc (87% yield; d.r. = 1.9:1).[4] α-Acetoxy ethers have also been utilized as Prins cyclization precursors in our studies toward the diastereoselective synthesis of polyfunctionalized tetrahydropyran ring systems.[5]

1. Department of Chemistry, University of California, Irvine, CA 92697-2025.

2. Pangborn, A. B.; Giardello, M. A.; Grubbs, R. H.; Rosen, R. K.; Timmers, F. J. *Organometallics* **1996**, *15*, 1518.

3. (a)Kopecky, D. J.; Rychnovsky, S. D. *J. Org. Chem.* **2000**, *65*, 191; (b) Dahanukar, V. H.; Rychnovsky, S. D. *J. Org. Chem.* **1996**, *61*, 8317.

4. Both reactions were performed in dichloromethane at -78°C.

5. (a) Rychnovsky, S. D.; Hu, Y. ; Ellsworth, B. *Tetrahedron Lett.* **1998**, *39*, 7271 (b) Rychnovsky, S. D.; Thomas, C. R. *Org. Lett.* **2000**, *2*, 1217 (c) Jaber, J. J.; Mitsui, K.; Rychnovsky, S. D. *J. Org. Chem.* **2001**, *66*, 4679.

Appendix

Chemical Abstracts Nomenclature (Collective Index Number); (Registry Number)

endo-1-Bornoxyethyl acetate: Ethanol, 1-[[(1R,2S,4R)-1,7,7-trimethylbicyclo[2.2.1]-h 2-yl]oxy]-, acetate (9); (284036-61-9)

(–)-Bornyl acetate: Bicyclo[2.2.1]heptan-2-ol, 1,7,7-trimethyl-, acetate, (1S,2R,4S)- (9); (5655-61-8)

Diisobutylaluminium hydride: Aluminum, hydrobis(2-methylpropyl)- (9); (1191-15-7)

Pyridine (8, 9); (110-86-1)

4-(Dimethylamino)pyridine: 4-Pyridinamine, N,N-dimethyl- (9); (1122-58-3)

Acetic anhydride: Acetic acid, anhydride (9); (108-24-7)

TABLE 1[3a]

DIBALH REDUCTIVE ACETYLATION OF VARIOUS ESTERS

Entry	Substrate	Product	Yield(%)
1	$(CH_3)_3C\text{-}C(=O)\text{-}O\text{-}CH_2\text{-}C(CH_3)_3$	$(CH_3)_3C\text{-}CH(OAc)\text{-}O\text{-}CH_2\text{-}C(CH_3)_3$	87
2	$Ph\text{-}CH_2CH_2\text{-}C(=O)\text{-}OC(CH_3)_3$	$Ph\text{-}CH_2CH_2\text{-}CH(OAc)\text{-}OC(CH_3)_3$	90
3	$Ph\text{-}CH_2CH_2\text{-}C(=O)\text{-}OCH_2Ph$	$Ph\text{-}CH_2CH_2\text{-}CH(OAc)\text{-}OCH_2Ph$	78
4	$ClCH_2\text{-}C(=O)\text{-}O\text{-}CH(CH_2CH_2Ph)(CH_2CH{=}CH_2)$	$ClCH_2\text{-}CH(OAc)\text{-}O\text{-}CH(CH_2CH_2Ph)(CH_2CH{=}CH_2)$	92 (d.r. = 1.1:1)
5	$BnO\text{-}CH_2CH_2\text{-}C(=O)\text{-}O(CH_2)_3CH_3$	$BnO\text{-}CH_2CH_2\text{-}CH(OAc)\text{-}O(CH_2)_3CH_3$	81
6	macrocyclic lactone (O, O)	macrocyclic acetal (OAc, O)	84
7	$F_3C\text{-}C(=O)\text{-}O\text{-}CH(CH_3)Ph$	$F_3C\text{-}CH(OAc)\text{-}O\text{-}CH(CH_3)Ph$	93 (d.r. = 2.2:1)

MILD AND SELECTIVE OXIDATION OF SULFUR COMPOUNDS IN TRIFLUOROETHANOL: DIPHENYL DISULFIDE AND METHYL PHENYL SULFOXIDE

(Disulfide, diphenyl- and Benzene, (methylsulfinyl)-)

1.1 eq 30% H_2O_2
Trifluoroethanol

1.1 eq 30% H_2O_2
Trifluoroethanol

Submitted by Kabayadi S. Ravikumar, Venkitasamy Kesavan, Benoit Crousse, Danièle Bonnet-Delpon, and Jean-Pierre Bégué.[1]

Checked by Jonathan Tripp and Dennis P. Curran.

1. Procedure

Diphenyl disulfide. In a 100-mL, round-bottomed flask equipped with a magnetic stirrer are placed 11.0 g (10.25 ml, 0.1 mol) of benzenethiol and 50 mL of trifluoroethanol (Note 1). The mixture is stirred and cooled in an ice bath (Note 2) and 12.5 mL (3.73 g, 0.11 mol) of 30% aqueous hydrogen peroxide (Note 3) is added dropwise over a period of 15 min through an addition funnel. After completion of the addition, the ice bath is removed and the reaction mixture is allowed to stir at room temperature for 24 hr. Diphenyl disulfide is sparingly soluble in trifluoroethanol and precipitates out of solution. Tho colids are collected on a Buchner funnel and dried under vacuum to afford 10.6 g of diphenyl disulfide (97%) (Note 4). Sodium sulfite 2.52 g (0.02 mol) is added to the mother liquor to decompose the excess peroxide and the mixture is heated in a water bath at 50°C for 30 min. A starch iodide test is negative. The liquid is transferred to a 100-mL, round-bottomed flask, fitted with a distillation unit having a Vigreux column (5 cm). The flask is heated in an oil bath and the solvent is distilled to recover the trifluoroethanol (Note 5).

Methyl phenyl sulfoxide. To a 100-mL, round-bottomed flask equipped with a magnetic stirrer are added 12.4 g (0.1 mol) of thioanisole and 50 mL of trifluoroethanol (Note 1). The mixture is stirred and cooled in an ice bath (Note 6) and 12.5 mL (3.73 g, 0.11 mol) of 30% aqueous hydrogen peroxide is added dropwise over 30 min using an addition funnel (Note 7). After completion of the addition, the ice bath is removed and the reaction mixture is stirred at room temperature for 8 hr (Note 8). Sodium sulfite (2.52 g, 0.02 mol) is added to decompose the excess hydrogen peroxide and the mixture is heated in a water bath at 50°C for 30 min (a starch iodide test is negative). The flask is fitted with a distillation unit having a Vigreux column (5 cm) and heated with an oil bath to remove the solvent by distillation (Note 5). Ether (100 mL) is added to the residue and the mixture is dried over 28 g of magnesium sulfate. Solids are removed by filtration and washed with ether (2 x 50 mL). The combined filtrate and washings are concentrated under vacuum to afford 14.9 g of a slightly yellow oil that crystallizes upon cooling in an ice bath. The crude sulfoxide is transferred to a 25-mL distillation flask with the aid of a small amount of dichloromethane. After removal of solvent, simple short path distillation (Note 9) of the crude product affords 12.8 g (91%) of pure methyl phenyl sulfoxide (Notes 10, 11).

2. Notes

1. Both benzenethiol and phenyl methyl sulfide are sparingly soluble in trifluoroethanol.

2. During the addition of hydrogen peroxide, the temperature of the reaction mixture increases. The submitters report that the addition can be done carefully without an ice bath and that no overoxidation is observed.

3. 30% Aqueous hydrogen peroxide is purchased from Fluka Chemical Company and handled with caution.

4. The product's physical and spectroscopic properties are: mp 52-53°C; ^{1}H NMR δ: 7.5-7.6 (m, 10 H); MS m/z 218 (M^+), 109.

5. Initially trifluoroethanol is obtained (37-38 mL, bp 77-79°C). Subsequently, a fraction (11-13 mL), containing a mixture of trifluoroethanol and water, is collected at 85-87°C. The ^{19}F NMR spectrum of this fraction was identical to that of trifluoroethanol. The recovered trifluoroethanol can be reused for the oxidation of sulfides.

6. When the addition of hydrogen peroxide is carried out at room temperature, the submitters reported that the reaction is exothermic and that 3-4% of the sulfone is formed.

7. When 2 equivalents of hydrogen peroxide are used, the submitters report that 0-2% of the sulfone is obtained. When phenyl methyl sulfoxide is oxidized under similar conditions, no sulfone is observed, even after 24 hr.

8. The submitters report that oxidation of phenyl methyl sulfide to phenyl methyl sulfoxide also can be achieved selectively by using hexafluoroisopropyl alcohol (HFIP) as the solvent; the reaction is complete after 10 min.

9. The technique of adding a small amount of activated carbon to the distillation pot afforded a colorless distillate.

10. A fore-fraction is not obtained.

11. The product's physical and spectroscopic properties are: bp 78-79°C (0.1 mm); ^1H NMR ($CDCl_3$) δ: 2.7 (s, 3 H); 7.5-7.7 (m, 5 H); MS m/z 140 (M^+), 125, 97, 77.

3. Discussion

Methods of preparation. Diphenyl disulfide can be prepared by the action of hydrogen peroxide in acidic or basic media,[2] iodine-hydrogen iodide,[3] Sm/$BiCl_3$ in water,[4b] $FeCl_3$/NaI,[4c] and by using bromine-aqueous potassium hydrogen carbonate.[4] Methyl phenyl sulfoxide has also been prepared from thioanisole by the action of hydrogen peroxide in methanol,[5] mCPBA,[5b] peroxytrifluoroacetic acid[6] and sodium periodate.[7]

Merits of preparation. Selective oxidative transformation of thiols to disulfides is of interest for both biological[8] and synthetic reasons.[2] Most of the existing methods involve the use of metal catalysts or halogenated reagents, which lead to the formation of side-products.

Organic sulfoxides are useful intermediates for the construction of many chemically and biologically significant molecules.[9] Most of the current methods for the oxidation of sulfides to sulfoxides[10] suffer from over-oxidation to sulfones. Even when controlling the reaction temperature, it is difficult to avoid over-oxidation in most of the reported procedures.

The present procedure is based on the use of fluoroalkyl alcohols as solvents in oxidation reactions. The method is efficient and versatile, and produces disulfides and sulfoxides under mild conditions. These reactions have also been developed using hexafluoroisopropyl alcohol (HFIP) as solvent with a large variety of substrates (alkyl sulfides, alkyl thiols, vinyl sulfides, fluorinated vinyl sulfides, thioglucosides) by Bégué et al.[11,12] Replacement of HFIP with trifluoroethanol, a more common and less expensive solvent, also allows the use of mild conditions and affords high yields of disulfides and sulfoxides without contamination. 30% Aqueous hydrogen peroxide is inexpensive and, since water is the sole byproduct, this method is environmentally friendly.

Cooling is advisable during the addition of hydrogen peroxide, but the remainder of the reaction can be carried out at ambient temperature. This method may be useful for large-scale reactions since trifluoroethanol can be recovered.

1. Laboratoire BIOCIS associé au CNRS, Centre d'Etudes Pharmaceutiques, Rue J.B. Clément, 92296 Chatenay-Malabry, France.
2. Capozzi, G.; Modena, G. In "Chemistry of the Thiol Group"; Patai, S., Ed.; John Wiley & Sons, New York, 1974; Part 2, p. 785.
3. Aida, T.; Akasaka, T.; Furukawa, N.; Oae, S. *Bull. Chem. Soc., Jpn.* **1976**, *49*, 1441.
4. (a) Drabowicz, J.; Mikolajczyk, M. *Synthesis* **1980**, 32; (b) Wu, X.; Rieke, R. D.; Zhu, L. *Synth. Commun.* **1996**, *26*, 191; (c) Iranpoor, N.; Zeynizadeh, B. *Synthesis* **1999**, *1*, 49.
5. (a) Drabowicz, J.; Lyzwa, P.; Popielarczyk M.; Mikolajczyk, M. *Synthesis* **1990**, 937; (b) Drabowicz, J.; Kielbasinski, P.; Mikolajczyk, M. In "Syntheses of Sulphones, Sulphoxides and Cyclic Sulphides"; Patai, S.; Rappoport, Z., Eds.; John Wiley and Sons: West Sussex, 1994; p. 109.
6. Venier, C. G.; Squires, T. G.; Chen, Y.-Y.; Hussmann, G. P.; Shei, J. C.; Smith, B. F. *J. Org. Chem.* **1982**, *47*, 3773.
7. Johnson, C. R.; Keiser, J. E. *Org. Synth., Coll. Vol. V* **1973**, 791.
8. Jocelyn, P. C. In "Biochemistry of the SH Group"; Academic Press: London, New York, 1972.
9. (a) Block, E. In "Reactions of Organosulfur Compounds"; Academic Press: New York, 1978; (b) Durst, T. In "Comprehensive Organic Chemistry"; Barton, D.; Ollis, W. D., Eds.; Pergamon: Oxford, 1979; Vol. 3.

10. Drabowicz, J.; Kielbasinski, P.; Mikolajczyk, M. In the "The Chemistry of Sulfones and Sulfoxides; Patai, S.; Rappoport, Z.; Stirling, C. J. M., Eds.; Wiley: New York, **1988**; p. 233.

11. (a) Ravikumar, K. S.; Bégué, J.-P.; Bonnet-Delpon, D. *Tetrahedron Lett.* **1998**, *39*, 3141; (b) Ravikumar, K. S.; Zhang, Y. M.; Bégué, J.-P.; Bonnet-Delpon, D. *Eur. J. Org. Chem.* **1998**, 2937; (c) Ravikumar, K. S.; Barbier, F.; Bégué, J.-P.; Bonnet-Delpon, D. *J. Fluorine Chem.* **1999**, *95*, 123.

12. Kesavan, V.; Bégué, J.-P.; Bonnet-Delpon, D. *Synthesis* **2000**, 223-225.

Appendix

Chemical Abstracts Nomenclature (Collective Index Number);

(Registry Number)

Benzenethiol (8,9); (108-98-5)

Diphenyl disulfide: Disulfide, diphenyl (9); (882-33-7)

Hydrogen peroxide (H_2O_2) (9); (7722-84-1)

Methyl phenyl sulfoxide: Benzene, (methylsulfinyl)- (9); (1193-82-4)

Thioanisole: Benzene, (methylthio)- (9); (100-68-5)

Trifluoroethanol: Ethanol, 2,2,2-trifluoro- (6,8,9); (75-89-8)

1-HYDROXY-3-PHENYL-2-PROPANONE

(2-Propanone, 1-hydroxy-3-phenyl-)

A. 1 → 2 (PrSH, KOH (aq), MeCN)

B. 2 → 3 (H_2SO_4, $EtOH/H_2O$)

Submitted by Marjorie See Waters, Kelley Snelgrove, and Peter Maligres.[1]

Checked by Günter Seidel and Alois Fürstner.

1. Procedure

Caution: The procedures should be carried out in a well-ventilated hood because of the extreme stench of the mercaptan products. All glassware used in the procedures should be rinsed with bleach solution prior to removal from the fume hood.

A. *3-Phenyl-2-(propylthio)-2-propen-1-ol* (**2**). A 250-mL, one-necked, round-bottomed flask, equipped with a magnetic stirring bar, thermocouple probe and nitrogen inlet, is charged with acetonitrile (150 mL), 3-phenyl-2-propyn-1-ol (19.8 g, 0.15 mol), 30 wt% KOH (2.84 g, 15.2 mmol) and 1-propanethiol (14.0 mL, 0.16 mol) (Note 1). The reaction mixture is heated to 60°C with a heating mantle and stirred for 6 hr (Note 2). After cooling to room temperature, the bulk of the solvent is removed under reduced pressure. The resulting yellow oil is taken up in 100 mL of tert-butyl methyl ether (TBME) and filtered through 50 g of silica gel (Note 3). The silica gel plug is washed with 300 mL of TBME. The filtrates are combined and concentrated under reduced pressure to give 30.83 g (99%) of 3-phenyl-2-(propylthio)-2-propen-1-ol as a pale yellow oil which can be used without further purification (Note 4).

B. *1-Hydroxy-3-phenyl-2-propanone* (**3**). A 1-L, three-necked, round- bottomed flask, equipped with a magnetic stirring bar, thermocouple probe, rubber septum and nitrogen inlet and outlet that leads to a scrubber containing bleach, is charged with 3-phenyl-2-(propylthio)-2-propen-1-ol (30.1 g, 0.14 mol), 150 mL of ethanol and 150 mL of 1M sulfuric acid. The suspension, which becomes a homogeneous yellow solution, is heated at 65°C with a heating mantle for 4 hr with vigorous stirring and a nitrogen sweep (Notes 5, 6).

After cooling to room temperature, the reaction mixture is transferred to a 1-L separatory funnel with the aid of 50 mL of water and washed with 250 mL of hexanes, which is discarded (Note 7). After adding 150 mL of brine to the aqueous phase, the product is extracted with ethyl acetate (3 x 250 mL). The combined organic extracts are dried over magnesium sulfate, filtered, and concentrated under reduced pressure (Note 8).

The residue is dissolved in 40 mL of ethyl acetate and filtered through a pad of Celite into a 500-mL round-bottomed flask. With stirring, 240 mL of n-heptane is added over 20 min to the ethyl acetate solution via a pressure-equalizing addition funnel. During the addition, the product begins to crystallize as a white solid. The volume of the mixture is reduced by distillation at reduced pressure (20°C, 10-20 mm) to about 150 mL. The mixture is then cooled to –5°C and aged for 1 hr. The solids are collected via vacuum filtration, washed with heptane (100 mL), and dried under vacuum at room temperature to afford 1-hydroxy-3-phenyl-2-propanone (18.9 g, 87% yield) as white crystals (Note 9).

2. Notes

1. 3-Phenyl-2-propyn-1-ol was purchased fom Lancaster Synthesis Ltd. and used as received. 1-Propanethiol was purchased from Aldrich Chemical Company, Inc., and used as received.

2. The reaction progress can be monitored by GC analysis (HP-1701, 15 m x 0.53 μm capillary column, 1.00 μm film thickness, 100°C increase at 15°/min to 250°C); the retention time was 4.6 min for the alkynol (**1**) and 7.9 min for the vinyl sulfide (**2**).

3. Silica gel is added to a 300 mL fritted glass funnel and pre-washed with TBME under vacuum, discarding the filtrate. The reaction mixture is poured onto the pad and filtered under vacuum, followed by the TBME washes. An orange layer of by-products remains at the top of the silica gel.

4. The crude product typically contains 2-3% of a by-product resulting from addition to the 3-position of the alkynol, 2-3% of the E-addition product, and residual propanethiol. Spectral data for **2** are as follows: IR cm^{-1}: 3358, 2962, 2930, 2871, 1598, 1491, 1445, 1378, 1291, 1237, 1098, 1078, 1009, 753, 694. ^{1}H NMR (400 MHz, $CDCl_3$) δ: 0.96 (t, 3H, J= 7.3), 1.60 (m, 2H), 2.76 (t, 2H, J = 7.3), 4.37 (s, 2H), 6.83 (s, 1H), 7.2-7.4 (m, 3H), 7.63 (d, 2H, J=7.6); ^{13}C NMR (100 MHz, $CDCl_3$) δ: 13.4, 23.4, 33.4, 67.0, 127.4, 128.1, 129.5, 129.9, 135.6, 136.3. The checkers obtained **2** in 95% yield. The checkers noted that compound **2** may eventually isomerize to 3-phenyl-2-(propylthio)propanal when kept in $CDCl_3$ which was not rigorously dried prior to use.

5. A moderate nitrogen purge is necessary to remove the propanethiol, which is formed in the reaction. Care should be taken to treat any distillate collected in the trap with bleach.

6. The reaction rate will depend on the stirring and nitrogen purge rates. The reaction should be monitored for completion (>98% conversion) by GC using the method described in Note 2; the retention time for hydroxyketone **3** is 4.9 min.

7. The hexanes wash removes thiol-containing by-products and should be treated with bleach prior to disposal.

8. The concentration is done at 20-25°C at the lowest possible pressure. The hydroxyketone (3) is reactive and all product-containing solutions should be maintained below 25°C during the work up.

9. The physical properties of **3** are as follows: mp 47-48°C; IR (KBr) cm^{-1}: 3378, 1721, 1496, 1454, 1402, 1291, 1076, 1043, 755, 701; ^{1}H NMR (400 MHz, $CDCl_3$) δ: 2.94 (s, 1H), 3.74 (s, 2H), 4.30 (s, 2H), 7.2-7.4 (m, 5H); ^{13}C NMR (100 MHz, $CDCl_3$) δ: 45.9, 67.7, 127.6, 129.0, 129.4, 132.8, 207.3. Anal. Calcd for $C_9H_{10}O_2$: C, 71.98; H, 6.71. Found: C, 71.79; H, 6.70. The checkers obtained material having mp 45-46°C in 83% yield.

Waste Disposal Information

All toxic materials were disposed of in accordance with "Prudent Practices in the Laboratory"; National Academy Press; Washington, DC, 1995.

3. Discussion

Hydroxyketones are versatile intermediates in the synthesis of pharmaceutical intermediates and heterocyclic molecules.[2] α-Aryl hydroxyketones have been prepared by reaction of aryl aldehydes with 1,4-dioxane followed by reduction with lithium aluminum hydride (LAH)[3] and by the selective LAH reduction of α-silyloxy α,β-unsaturated esters.[4] Wissner[5] has shown that treatment of acid chlorides with tris(trimethylsilyloxy)ethylene affords alkyl and aryl hydroxymethyl ketones. 1-Hydroxy-3-phenyl-2-propanone (3) has been generated by the osmium-catalyzed oxidation of phenylpropene[6] and by the palladium-catalyzed rearrangement of phenyl epoxy alcohol[7] both in 62% yield.

The present method is superior in that it requires neither oxidation nor reduction and does not use expensive and toxic heavy metal catalysts. The two step process offers mild conditions and high yields. Recovery of the thiol is possible by using fractional distillation during the hydrolysis, making the process efficient on large scale. The procedure is general for a variety of α-aryl hydroxyketones.[8]

1. Department of Process Research, Merck Research Laboratories, Merck & Co., Inc., Rahway, NJ 07065
2. (a) Horikawa, M.; Nakajima, T.; Ohfune, Y. *Synlett* **1997**, 253-254; (b) Liu, K.; Wood, H. B.; Jones, A. B. *Tetrahedron Lett.* **1999**, *40*, 5119-5122; (c) Jerris, P. J.; Wovkulich, P. M.; Smith, A. B., III *Tetrahedron Lett.* **1979**, *20*, 4517-4520.
3. Fetizon, M.; Hanna, I.; Rens, J. *Tetrahedron Lett*. **1985**, *26*, 3453-3456.
4. Dalla, V.; Catteau, J. P *Tetrahedron* **1999**, *55*, 6497-6510.
5. Wissner, A. *Tetrahedron Lett.* **1978**, *19*, 2749-2752.
6. Murahashi, S.; Naota, T.; Hanaoka, H. *Chem. Lett*. **1993**, *10*, 1767-1770.
7. Vankar, Y. D.; Chaudhuri, N. C.; Singh, S. P. *Synth. Comm*. **1986**, *16*, 1621-1626.
8. Waters, M. S.; Cowen, J. A.; McWilliams, J. C.; Maligres, P. E.; Askin, D. *Tetrahedron Lett*. **2000**, *41*, 141-144.

Appendix

Chemical Abstracts Nomenclature (Collective Index Number);

(Registry Number)

1-Hydroxy-3-phenyl-2-propanone: 2-Propanone, 1-hydroxy-3-phenyl- (8, 9); (4982-08-5)

3-Phenyl-2-propyn-1-ol: 2-Propyn-1-ol, 3-phenyl- (8, 9); (1504-58-1)

Propanethiol (9); (79869-58-2)

DIETHYLAMINOTRIMETHYLSILANE-CATALYZED
1,4-ADDITION OF ALDEHYDES TO VINYL KETONES:
(3R)-3,7-DIMETHYL-2-(3-OXOBUTYL)-6-OCTENAL
[6-Octenal, 3,7-dimethyl-2-(3-oxobutyl)-, (3R)-]

CHO + O —Et_2NSiMe_3→ CHO, O, Me

Submitted by Hisahiro Hagiwara, Hiroki Ono, and Takashi Hoshi.[1]
Checked by Kwame Nti-Addae and Steven Wolff.

1. Procedure

(3R)-3,7-Dimethyl-2-(3-oxobutyl)-6-octenal. To a two-necked, 300-mL flask, equipped with a magnetic stirrer, addition funnel, and condenser with a nitrogen balloon on the top, and immersed in an ice-cooled bath is added 18.1 mL of citronellal (100 mmol) (Note 1), 12.5 mL of 3-buten-2-one (150 mmol) (Note 2), 1.9 mL of diethylaminotrimethylsilane (10 mmol) (Notes 3, 4) and 400 mL of acetonitrile (Note 5). Following the addition, the resulting homogeneous solution is heated at reflux in an oil bath and the progress of the reaction is monitored by TLC analysis of the product using Merck silica gel 60 F_{254} plates: R_f = 0.39 (5:1 hexane:ethyl acetate). After 46 hr, when the citronellal has been consumed, the resulting solution is concentrated to dryness under reduced pressure and the residue is distilled using a Kugelrohr apparatus twice (oven temperature 120~140°C at 1.9 mm) to give (3R)-3,7-dimethyl-2-(3-oxobutyl)-6-octenal (21.6 g, 96%) (Notes 6-8) as a pale yellow liquid.

2. Notes

1. R-(+)-Citronellal was obtained from Tokyo Kasei Kogyo Co. and used as received.

2. 3-Buten-2-one was obtained from Merck & Co., Inc. and used as received.

3. Diethylaminotrimethylsilane was obtained from Shin-Etsu Chemical Co., Ltd. and used as received.

4. Diethylaminotrimethylsilane can be prepared according to the procedure by Middleton, W. J.; Bingham, E. M. *Org. Synth., Coll. Vol. VI*, **1988**, 440.

5. Acetonitrile (Kanto Chemical Co., Ltd. special grade) was used as received.

6. The purity of the product was determined to be 100% by medium pressure LC (silica gel packed column, eluent: 5:1 n-hexane:ethyl acetate).

7. The physical properties of (3R)-3,7-dimethyl-2-(3-oxobutyl)-6-octenal, a 1:1 mixture of diastereoisomers, are as follows: IR (thin film) cm^{-1}: 2930, 2709, 1722, 1450, 1376, 1240 and 1164; ^1H NMR ($CDCl_3$, 200 MHz) δ: 0.89 (d, 1.5 H, J = 6.9), 0.99 (d, 1.5 H, J = 6.9), 1.60 (s, 3 H), 1.69 (s, 3 H), 1.08-2.10 (m, 7 H), 2.13 (s, 3 H), 2.26-2.65 (m, 3 H), 5.08 (m, 1 H), 9.60 (d, 0.5 H, J = 2.4) and 9.64 (0.5 H, J = 2.9); ^{13}C (50 MHz) δ: 15.9, 16.7, 17.6, 18.0, 19.7, 25.1, 25.6, 29.9, 32.2, 33.3, 33.8, 34.4, 41.36, 41.42, 55.9, 56.2, 123.7, 131.9, 204.9, 205.2, 207.9 and 208.0; m/z 224 (M^+, 0.1%), 148 (32), 109 (26), 95 (38), 82 (32), 71 (25), 69 (52), 58 (28), 55 (37), 43 (100) and 41 (68); Calcd for $C_{14}H_{24}O_2$; M^+, 224.1776. Found: M^+, 224.1798.

8. When 0.5 eq of Et_2NTMS was used, the reaction is complete in 19 hr in 100% yield as shown in the Table, entry 8.

Waste Disposal Information

All toxic materials were disposed of in accordance with "Prudent Practices in the Laboratory"; National Academy Press; Washington, DC, 1995.

3. Discussion

This is a mild, simple and practical procedure for 1,4-addition of an aldehyde to methyl vinyl ketone,[2] without converting the aldehyde into an enamine or a silyl enol ether. The products, substituted 5-ketoaldehydes, are important compounds, especially for the preparation of substituted 2-cyclohexen-1-one derivatives, which have been versatile starting materials for syntheses of natural products such as terpenoids.[3] These 5-ketoaldehydes have been prepared previously by the 1,4-addition of modified aldehydes, i.e., morpholinoenamines of aldehydes,[4,5] trimethylsilyl enol ethers of aldehydes in the presence of a Lewis acid,[6] or diethylallylamine in the presence of a catalytic amount of a Ru complex,[7] to methyl vinyl ketones.

The reaction is carried out simply by refluxing an acetonitrile solution of the aldehyde, vinyl ketone and 10 mol% of diethylaminotrimethylsilane until disappearance of the aldehyde (monitored by TLC analysis). Bulb-to-bulb distillation provides the 5-ketoaldehydes without aqueous work up. When the amount of diethylaminotrimethylsilane is increased, the reaction proceeds faster (Table, entry 8), although the resulting reaction mixture is not colorless. Some representative examples of the present reaction are shown in the Table. The mildness of the reaction is well exemplified by obtaining satisfactory yields in the reactions with acid- or base-sensitive aldehydes (having a THP or acetyl protecting group) (Table, entries 4 and 5).

The reaction also can be carried out without solvent, although the yields using isobutyraldehyde and citronellal were relatively low.

1. Graduate School of Science and Technology, Niigata University, 8050, 2-nocho, Ikarashi, Niigata 950-2181, Japan.
2. Hagiwara, H.; Kato, M. *Tetrahedron Lett.* **1996**, *37*, 5139; Hagiwara, H.; Komatsubara, N.; Ono, H.; Okabe, T.; Hoshi, T.; Suzuki, T.; Ando, M.; Kato, M. *J. Chem. Soc., Perkin Trans. 1* **2001**, 316.
3. Hagiwara, H.; Okabe, T.; Ono, H.; Kamat, V. P.; Hoshi, T.; Suzuki, T.; Ando, M. *J. Chem. Soc., Perkin Trans. 1* **2002**, 895.
4. Stork, G.; Brizzolara, A.; Landesman, H.; Szmuszkovicz, J.; Terrell, R. *J. Am. Chem. Soc.* **1963**, *85*, 207.
5. Brown, M. *J. Org. Chem.* **1968**, *33*, 162.
6. Duhamel, P.; Hennequin, L.; Poirier, J. M.; Tavel, G.; Vottero, C. *Tetrahedron* **1986**, *42*, 4777.
7. Nakajima, T.; Maruyama, Y.; Shimizu, I. In "Abstracts of Papers of 65th Annual Meeting of Chemical Society of Japan", Tokyo, 1993; Vol. II, p. 335.

Appendix

Chemical Abstracts Nomenclature (Collective Index Number); (Registry Number)

(3R)-3,7-Dimethyl-2-(2-oxobutyl)-6-octenal: 6-Octenal, 3,7-dimethyl-2-(3-oxobutyl)-, (3R)- (9); (131308-24-2)

R-(+)-Citronellal: 6-Octenal, 3,7-dimethyl-, (3R)- (9); (2385-77-5)

Methyl vinyl ketone: 3-Buten-2-one (8,9); (78-94-4)

Diethylaminotrimethylsilane: Silanamine, N,N-diethyl-1,1,1-trimethyl- (9); (996-50-9

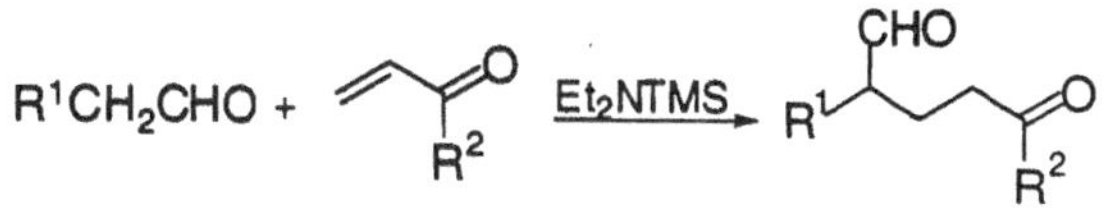

Table 1,4-Addition of aldehydes with vinyl ketones

Entry	Aldehyde (R^1)	Vinyl ketone (R^2)	Time (hr)	Product	Isolated yield (%)
1	C_8H_{17}	Me	20	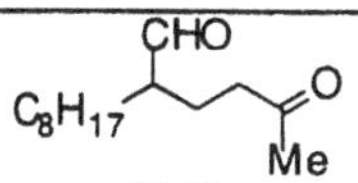	78
2	i-Pr	Me	9	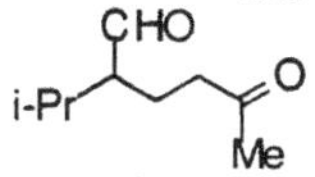	64
3	$PhCH_2$	Me	6	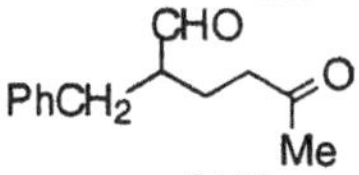	59
4	$THPOC_3H_6$	Me	6	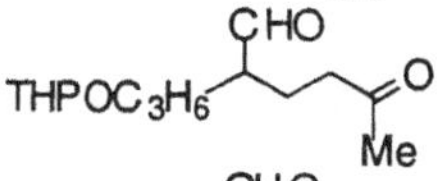	89
5	$AcOC_8H_{16}$	Me	9	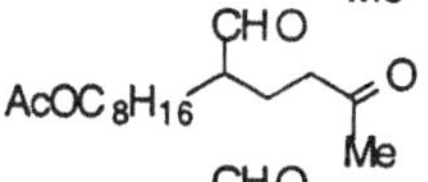	76
6	C_8H_{17}	Et	10.5	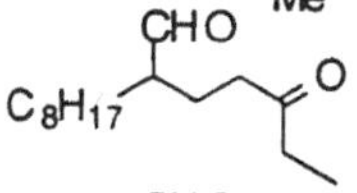	76
7	$PhCH_2$	Et	4	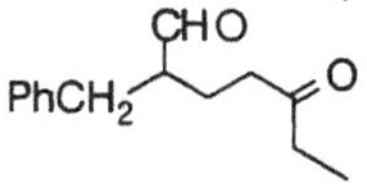	68
8[a]	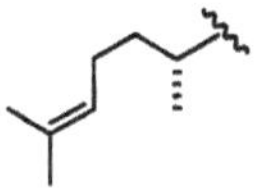	Me	19	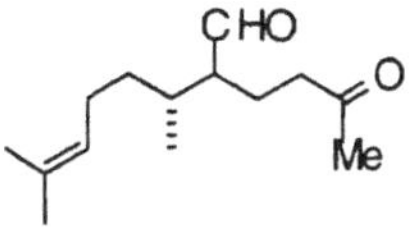	100

[a]Et_2NTMS (0.5 equiv.) was used.

PREPARATION OF 2-CHLORO-1,3-BIS(DIMETHYLAMINO)TRIMETHINIUM HEXAFLUOROPHOSPHATE

[Methanaminium, N-[2-chloro-3-(dimethylamino)-2-propenylidene]-N-methyl-, hexafluorophosphate(1-)]

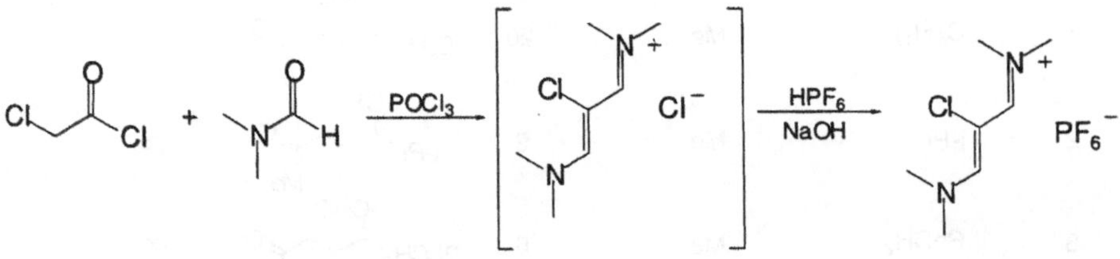

Submitted by Ian W. Davies, Jean-Francois Marcoux, and Jeremy Taylor.[1]

Checked by Yu Fan and Scott E. Denmark.

1. Procedure

A 250-mL, three-necked, round-bottomed flask, equipped with a stir bar, nitrogen bubbler (Note 1), thermocouple, and pressure-equalizing addition funnel, is charged with dimethylformamide (DMF) (60 mL) (Note 2). Chloroacetyl chloride (14.13 g, 0.125 mol) (Notes 3-5) is added over 5 min at 50°C. The mixture is then heated to 65–70°C to give a clear, yellow solution. Phosphorus oxychloride (19.20 g, 0.125 mol) (Notes 6, 7) is added at ~5 mL/hr maintaining the temperature of the mixture at 70°C. After completion of the addition, the mixture is heated for 3 hr to give a deep red solution. The reaction mixture is cooled to 20–30°C and transferred to a pressure-equalizing addition funnel. The reaction flask is rinsed with DMF (10 mL), which is added to the addition funnel. A solution of sodium hexafluorophosphate is prepared by the addition of 5N aqueous sodium hydroxide (NaOH) solution (37 mL) to a solution of hexafluorophosphoric acid *(Caution: Hexafluorophosphoric acid is toxic and corrosive)* (33.15 g of 60 wt%, 0.135 mol) in water (150 mL) in a 1-L, three-necked, round-bottomed flask fitted with a mechanical stirrer (Note 8). The vinamidinium chloride solution (~80 mL) and 5N aqueous NaOH solution (70 mL) are added concurrently over 1 hr to the solution of sodium hexafluorophosphate in water (150 mL) with stirring at a temperature of <10°C to give a yellow slurry (Note 9). The mixture is aged for 1 hr, then the solids are collected under vacuum. The crude solids are washed with water (75 mL) (Note 10), then transferred to a 500-mL, three-necked, round-bottomed flask equipped with a mechanical stirrer and thermocouple. Water (225 mL) and 2-propanol (60 mL) are added to the flask. The mixture is heated to 70°C to dissolve the solids (Note 11). The mixture is then cooled to 0–5°C over 1 hr. The light-yellow solids are collected by filtration, washed with cold water/2-propanol (50 mL, 20/1), and dried under reduced pressure at <40°C (Note 12) to give 2-chloro-1,3-bis(dimethylamino)trimethinium hexafluorophosphate as a light-yellow solid (29.1 g, 77%) (Notes 13, 14). *Caution: The salt is a skin and eye irritant and was positive in the Ames mutagenesis assay. It should be handled in a hood with adequate personal protective equipment.* The product is >99.9% pure by HPLC analysis (Note 15).

2. Notes

1. The reaction is run under dry nitrogen.

2. A fresh bottle of anhydrous DMF from Aldrich Chemical Co., Inc. was used. The water content of the DMF was 0.024 g/L by Karl Fischer titration.

3. The checkers purchased chloroacetyl chloride from Aldrich Chemical Co., Inc.

4. Chloroacetic acid may be used in place of chloroacetyl chloride by adjusting the charge of phosphorus oxychloride to 0.25 mol. The submitters purchased chloroacetic

acid from Säurefabrik Schweizerhall (CH). The material is also available from Aldrich Chemical Co., Inc.

5. The addition is mildly exothermic.

6. A syringe pump was used for this addition by the checkers.

7. When chloroacetic acid is used in place of chloroacetyl chloride, the charge of phosphorus oxychloride is increased to 0.25 mol.

8. The checkers purchased hexafluorophosphoric acid from Aldrich Chemical Co., Inc.

9. The addition is mildly exothermic. Sodium hexafluorophosphate is commercially available although significantly more expensive. The sodium salt may be used in place of the hexafluorophosphoric acid by omitting the initial sodium hydroxide charge.

10. The crude solids may be dried and were ~98 area% pure by HPLC analysis according to the submitters. The crude material may be used in a number of applications without recrystallization.

11. The submitters have observed that aging at elevated temperature for long periods (>8 hr) will lead to a 3–5% decrease in yield. Over-heating the solids on the walls of the flask may discolor the product.

12. According to the submitters there is a minor exotherm initiating at 60°C by differential scanning (Reactive System Screening Tool) (dT/dt °C/min 2.0, dP/dt °C/min <1). The exotherm is non-hazardous but will lead to deterioration of product quality presumably due to hydrolysis.

13. The submitters have run this procedure at various scales up to 300 kg with yields ranging from 73-79%.

14. The product is fully characterized: mp 125-126°C; ^{1}H NMR (400 MHz DMF-d_7) δ 3.44 (s, 6 H), 3.60 (s, 6 H), 7.97 (s, 2 H); ^{13}C NMR (125 MHz DMF-d_7) δ 39.9, 49.7, 92.6, 161.1; ^{19}F NMR (470 MHz DMF-d_7) δ -72.3 (d, J = 708); ^{31}P NMR (202 MHz DMF-d_7) 143.2 (septet, J = 709); Anal. Calcd. for $C_7H_{14}ClN_2PF_6$: C, 27.42; H, 4.60; N, 9.14;

Cl, 11.56; F, 37.18; P, 10.10. Found: C, 27.09; H, 4.86; N, 8.85; Cl, 11.47; F, 37.14; P, 10.09%. IR (KBr) cm^{-1}: 1610(s), 1492(w), 1424(s), 1404(s), 1291(s), 1200(s), 1117(s), 1045(m), 963(w), 840(s), 740(w); MS (FAB 70 eV): 161(100), 162(3), 163(32), 164(3). X-ray crystallographic analysis reveals a "W"-conformation.

15. A reverse phase ion-pairing HPLC method was developed by the submitters for analysis. Chromatographic conditions: A 10-µL sample (0.1 mg/mL in acetonitrile) is injected onto a suitable liquid chromatograph equipped with a Waters Symmetry Shield RP18 column, 250 x 4.6 mm, 5 µm particle size at 40°C with a mobile phase of 0.404 g/L heptanesulfonic acid, sodium salt + 0.1% phosphoric acid (Component A, pH 2.2) and acetonitrile (Component B) at a flow rate of 1.0 mL/min, programmed with a linear gradient from 95:5 A:B (v/v) to 30:70 A:B (v/v) over 20 min. Detection is achieved by UV at 300 nm. The retention time is approximately 10 min.

Waste Disposal Information

All toxic materials were disposed of in accordance with "Prudent Practices in the Laboratory"; National Academy Press, Washington, DC, 1995.

3. Discussion

Vinamidinium salts are important intermediates in the synthesis of heterocycles.[2] The 2-chloro-1,3-bis(dimethylamino)trimethinium hexafluorophosphate salt has been used in the preparation of the highly selective Cox-2 inhibitor etoricoxib (Scheme 1).[3] This method describes a straight-forward preparation of the hexafluorophosphate salt which is a crystalline, thermally and shock-stable, non-hygroscopic solid. The submitters have extensively studied the preparation of vinamidinium salts and demonstrated that the method is applicable to substituted acetic acids that contain an electron-withdrawing group (Table 1).[4] The annulation reaction is also general and useful for the preparation of pyridines, pyridones and pyridine N-oxides.[5]

Scheme 1

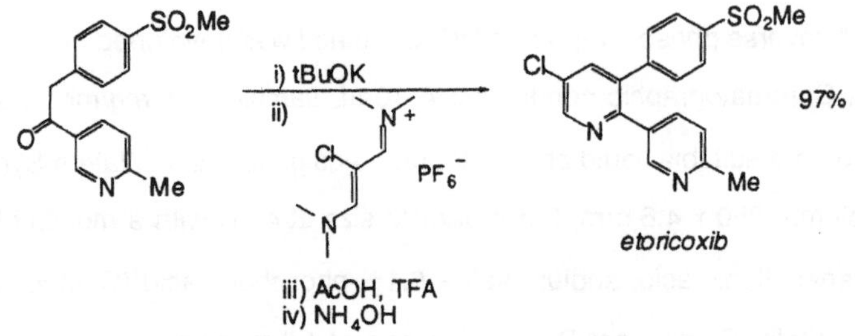

SO_2Me · O · N · Me

i) tBuOK
ii) Cl, N^+, N, PF_6^-
iii) AcOH, TFA
iv) NH_4OH

Cl · SO_2Me · N · N · Me · 97%

etoricoxib

1. Department of Process Research, Merck & Co., Inc., PO Box 2000, Rahway, NJ 07065-0900.
2. (a) For examples see: Gupton, J. T.; Petrich, S. A.; Hicks, F. A.; Wilkinson, D. R.; Vargas, M.; Hosein, K. N.; Sikorski, J. A. *Heterocycles* **1998**, *47*, 689. (b) Kase, K.; Katayama, M.; Ishihara, T.; Yamanaka, H.; Gupton, J. T. *J. Fluor. Chem.* **1998**, *90*, 29.
3. Davies, I. W.; Marcoux, J. -F.; Corley, E. G.; Journet, M.; Cai, D.-W.; Palucki, M.; Wu, J.; Larsen, R. D.; Rossen, K.; Pye, P. J.; DiMichele, L.; Dormer, P.; Reider, P. J. *J. Org. Chem.* **2000**, *65*, 8415.
4. (a) Davies, I. W.; Marcoux, J.-F.; Wu, J.; Palucki, M.; Corley, E. G.; Robbins, M. A.; Tsou, N.; Ball, R. G.; Dormer, P.; Larsen, R. D.; Reider, P. J. *Org. Chem.* **2000**, *65*, 4571. (b) The method is general but the reaction times/temperatures may require some slight modification to give optimal yields of the desired vinamidinium. In the case of the trifluoromethylvinamidinium 60°C is optimal; Davies, I. W.; Tellers, D. M.; Shultz, C. S.; Fleitz, F. J.; Cai, D. -W.; Sun, Y. *Org. Lett.* **2002**, *4*, 2969.

5. (a) Marcoux, J.-F.; Corley, E. G.; Rossen, K.; Pye, P.; Wu, J.; Robbins, M. A.; Davies, I. W.; Larsen, R. D.; Reider, P. J. *Org. Lett.* **2000**, *2*, 2339. (b) Marcoux, J.-F.; Marcotte, F.-A.; Wu, J.; Dormer, P. G.; Davies, I. W.; Hughes, D.; Reider, P. J. *J. Org. Chem.* **2001**, *66*, 4194. (c) Davies, I. W.; Marcoux, J.-F.; Reider, P. J. *Org. Lett.* **2001**, *3*, 209.

Table 1

Entry	R (vinamidinium PF_6^- salt)	Yield
1	R = Cl	78 %
2	R = Br	78 %
3	R = I	60 %
4	R = CF_3	68 %
5	R = Ph	90 %
6	R = 4-NO_2-C_6H_4-	80 %
7	R = 4-MeO-C_6H_4-	75 %
8	R = (phthalimido, N–)	68 %

Appendix

Chemical Abstracts Nomenclature (Collective Index Number); (Registry Number)

2-Chloro-1,3-bis(dimethylamino)trimethinium hexafluorophosphate: Methanaminium, N-[2-chloro-3-(dimethylamino)-2-propenylidene]-N-methyl-, hexafluorophosphate(1-) (9); (249561-98-6)

Chloroacetyl chloride: Acetyl chloride, chloro- (8, 9); (79-04-9)

Dimethylformamide: Formamide, N,N-dimethyl- (8, 9); (68-12-2)

Phosphorus oxychloride: Phosphoric trichloride (9); (10025-87-3)

Hexafluorophosphoric acid: Phosphate (1-), hexafluoro-, hydrogen (8, 9); (16940-81-1)

TRANSFORMATION OF PRIMARY AMINES TO N-MONOALKYLHYDROXYLAMINES: N-HYDROXY-(S)-1-PHENYLETHYLAMINE OXALATE

(Benzenemethanamine, N-hydroxy-α-methyl-, (S)-, ethanedioate)

A. (S)-1-phenylethylamine $\xrightarrow[CH_3CN]{BrCH_2CN,\ i\text{-}Pr_2NEt}$ (S)-[(1-phenylethyl)amino]acetonitrile

B. (S)-[(1-phenylethyl)amino]acetonitrile $\xrightarrow[CH_2Cl_2]{MCPBA}$ nitrone (N^+–O^-, =CH–CN)

C. nitrone $\xrightarrow[(COOH)_2,\ MeOH]{NH_2OH \cdot HCl}$ N-hydroxy-(S)-1-phenylethylamine $\cdot (COOH)_2$

Submitted by Hidetoshi Tokuyama, Takeshi Kuboyama, and Tohru Fukuyama.[1]

Checked by Scott E. Denmark and Jeromy J. Cottell.

1. Procedure

Caution! Hydroxylamines may cause explosions under certain conditions. Careful handling and proper safety equipment are needed especially when they are heated or have not been converted to salts with acids.

A. *(S)-[(1-Phenylethyl)amino]acetonitrile.* A 1-L, round-bottomed flask equipped with a Teflon-coated magnetic stirrer bar, rubber septum and argon gas inlet is charged with (S)-1-phenylethylamine (10.17 g, 83.92 mmol) (Note 1), acetonitrile (MeCN) (170 mL) (Note 2), and diisopropylethylamine (i-Pr_2NEt) (29.2 mL, 168 mmol) (Note 3). After stirring the solution for 5 min, bromoacetonitrile (6.43 mL, 92.3 mmol) (Note 3) is added via syringe over 10 min. The reaction mixture is stirred at ambient temperature until completion of the reaction (Note 4). The mixture is then concentrated on a rotary evaporator to give a white solid, to which is added saturated aqueous sodium bicarbonate ($NaHCO_3$) solution (100 mL); the suspension is extracted with dichloromethane (CH_2Cl_2) (100 mL). The organic phase is washed with brine (100 mL), and the combined aqueous phases are back-extracted with CH_2Cl_2 (3 x 100 mL). The combined organic extracts are dried over anhydrous sodium sulfate, filtered, and concentrated under reduced pressure to give a slightly yellow oil (21.6 g). To the crude oil is added CH_2Cl_2 (10 mL) and the resulting solution is passed through a short column of silica gel (5 cm i. d. x 10 cm, 100 g of SiO_2) (Note 5). A forerun was obtained by elution with 400 mL of 9:1 hexanes:ethyl acetate, followed by collection of the desired fractions with 100 mL of 9:1 hexanes:ethyl acetate, then 700 mL of 7:3 hexanes:ethyl acetate (Note 6). Concentration of the appropriate fractions on a rotary evaporator and then under high vacuum gives (S)-[(1-phenylethyl)amino]acetonitrile as a colorless oil (13.62 g, 101%) (Note 7).

B. *[(1S)-1-Phenylethyl]imino]acetonitrile N-oxide*. A 1-L, three-necked flask equipped with a mechanical stirrer, argon gas inlet, and thermometer is charged with (S)-[(1-phenylethyl)amino]acetonitrile (10.34 g, 64.54 mmol) and CH_2Cl_2 (200 mL) (Note 8), and the resulting solution is cooled in an ice bath. m-Chloroperbenzoic acid (MCPBA) (ca. 77%, 34.3 g, 153 mmol) (Note 9) is added in portions over 30 min (Notes 10-12). After completion of the addition, the ice bath is removed and the mixture is stirred at ambient temperature. When the reaction is complete (Note 13), the mixture is cooled in an ice bath again, and aqueous sodium thiosulfate ($Na_2S_2O_3$) (16.0 g, 129 mmol, in 60 mL of H_2O) and saturated $NaHCO_3$ (200 mL) are added. The resulting slurry is stirred vigorously for 15 min until the white solid completely dissolves. The two-phase solution is separated using a separatory funnel. The aqueous phase is extracted with CH_2Cl_2 (100 mL). The organic layer is washed with brine (100 mL), and the combined aqueous phases are back-extracted with CH_2Cl_2 (3 x 100 mL). The combined organic extracts are dried over anhydrous sodium sulfate, filtered, and concentrated on a rotary evaporator to afford crude nitrone (11.5 g) as slightly yellow crystals (Note 14).

C. *N-Hydroxy-(S)-1-phenylethylamine oxalate.* A 1-L, round-bottomed flask equipped with a reflux condenser, Teflon-coated magnetic stirrer bar, and argon gas inlet is charged with crude nitrone (11.5 g) and methanol (MeOH) (130 mL) (Note 15). After addition of hydroxylamine hydrochloride (20.8 g, 323 mmol) (Note 16) in one portion at ambient temperature, the mixture is warmed to 60°C (55°C internal) and is stirred at that temperature for 2 hr (Note 17). The reaction mixture is cooled to room temperature and diluted with CH_2Cl_2 (260 mL). After stirring for 5 min, the resulting precipitate is collected by filtration and the filter cake is washed with CH_2Cl_2 (40 mL). The filtrate is neutralized with saturated $NaHCO_3$ (150 mL) and partitioned. The aqueous phase is extracted with CH_2Cl_2 (100 mL). The organic phase is washed with brine (100 mL), and the combined aqueous phases are back-extracted with CH_2Cl_2 (3 x 100 mL). The combined organic extracts are dried over anhydrous sodium sulfate, filtered and concentrated to a minimum volume at a temperature below 25°C (Note 18) under reduced pressure. To the residue is added a MeOH (25 mL) solution of oxalic acid (11.6 g, 129 mmol) (Note 19) and ether (Et_2O) (100 mL), precipitating white crystals, which are collected by filtration and washed with Et_2O (3 x 20 mL). After drying under reduced pressure, (S)-N-1- phenylethylhydroxylamine oxalate (12.6 g, 86% for two steps) is obtained as white crystals. The oxalate salt was recrystallized from hot ethanol (110 mL) and washed with Et_2O (50 mL) to provide analytically pure material (10.45, 71% for two steps) (Notes 20, 21). Concentration and recrystallization of the mother liquor provides additional (S)-N-1-phenylethylhydroxylamine oxalate (0.85 g, 6%)

2. Notes

1. (S)-1-Phenylethylamine was purchased from Aldrich Chemical Co., Inc. and used as received. The enantiomeric purity of (S)-1-phenylethylamine was determined as the dinitrobenzoyl derivative to be 95% ee using a chiral stationary phase HPLC column and a chiral stationary phase SFC column (see Note 21). The submitters used reverse phase CSP HPLC analysis for this purpose (Daicel CROWNPAK CR, aq. $HClO_4$ (pH 1.5), 0.8 mL/min, 25°C, 210 nm).

2. Reagent grade acetonitrile was purchased from Fisher Scientific Co. and used as received.

3. Diisopropylethylamine and bromoacetonitrile were purchased from Aldrich Chemical Co., Inc. and used as received.

4. There is a slight exotherm (~10°C) over the first 30 min. The reaction typically takes 10-12 hr to complete and was monitored by TLC analysis on Merck silica gel 60 F-254 plates eluting with isopropylamine:hexane:EtOAc (5:45:50), R_f = 0.68 (visualized with 254-nm UV lamp and ninhydrin spray).

5. Silica gel was purchased from Merck (40-100 mesh, spherical, neutral).

6. The submitters found that the crude product can be purified by distillation (110-120°C, 1.5 mm). Starting from 24.3 g of amine, 21.8 g (68%) of the desired product was isolated. The reduced yield was due to dialkylation of the product with unreacted bromoacetonitrile during distillation.

7. While the product was sufficiently pure for the subsequent step, ^1H NMR analysis indicated the presence of 4 wt.% of bromoacetonitrile, which was absent after the Step B work-up. To secure analytically pure material, the checkers distilled

chromatographically purified product to obtain 9.93 g (74%) of (S)-[(1-phenylethyl)-amino]acetonitrile. The pure product exhibits the following physical properties: bp 91-93°C (1 mm); $[\alpha]_D^{27}$ −217 (c 1.23, $CHCl_3$), $[\alpha]_D^{23}$ −199.3 (c 1.08, EtOH), lit.[2] $[\alpha]_D^{25}$ −248.8 (c 4.7, benzene); IR (film) cm^{-1}: 3335, 2968, 2235, 1957, 1887, 1818, 1494, 1452, 1207, 1132, 870, 763; ^1H NMR (500 MHz, $CDCl_3$) δ: 1.42 (d, J = 6.4, 3 H), 1.67 (s (br), 1H), 3.28 (d, J = 17.6, 1 H), 3.58 (d, J = 17.6, 1 H), 4.04 (q, J = 6.5, 1 H), 7.26-7.34 (m, 5H); ^{13}C NMR (166 MHz, $CDCl_3$) δ: 23.9, 35.0, 56.7, 117.8, 126.9, 127.7, 128.7, 142.8. Anal. Calcd for $C_{10}H_{12}N_2$: C, 74.97; H, 7.55; N, 17.48. Found: C, 74.94; H, 7.70; N, 17.32.

8. Reagent grade dichloromethane was purchased from Fisher Scientific Co. and used as received.

9. MCPBA (ca. 65%) was purchased from Aldrich Chemical Co., Inc. and used as received.

10. The reaction temperature was maintained below 7°C.

11. Alternatively, conversion to the nitrone could be carried out with a catalytic amount of sodium tungstate and hydrogen peroxide[3] as described below: A 200-mL, round-bottomed flask equipped with a 30-mL, dropping funnel, Teflon-coated magnetic stirbar and argon gas inlet is charged with (S)-[(1-phenylethyl)amino]acetonitrile (6.10 g, 38.1 mmol), MeOH (64 mL), and sodium tungstate dihydrate ($Na_2WO_4 \cdot 2H_2O$) (504 mg, 1.53 mmol) in one portion. With cooling with an ice-bath, 30% aqueous hydrogen peroxide (14.7 mL, 153 mmol) is added to the solution over 20 min. The reaction mixture is then allowed to warm to ambient temperature. After TLC analysis shows completion of the reaction (usually 10 to 20 hr), aqueous $Na_2S_2O_3$ (15 mL) is added slowly with cooling (ice-bath). The resulting suspension is extracted with CH_2Cl_2 (100 mL). The organic extracts are washed with brine (100 mL), and the combined aqueous phases are back-extracted with CH_2Cl_2 (3 x 100 mL). The combined extracts are dried over anhydrous sodium sulfate, filtered, and concentrated to dryness on a rotary evaporator to give crude nitrone (6.1 g) as a slightly yellow solid. The nitrone is directly

converted to hydroxylamine oxalate without further purification (5.40 g, 62% yield for 2 steps). The enantiomeric purity of the hydroxylamine was shown to be >99% ee upon reduction with Zn/AcOH and derivatization (see Note 21).

12. Conversion to the nitrone could also be carried out with magnesium monoperoxyphthalate (MMPP)[4] as follows: A 1-L, three-necked, flask equipped with a 30-mL addition funnel, Teflon-coated magnetic stirbar, thermometer and argon gas inlet, is charged with an aqueous solution (153 mL) of magnesium monoperoxyphthalate hexahydrate (26.0 g, 80%, 42 mmol), which is then cooled in an ice-bath. A solution of (S)-[(1-phenylethyl)amino]acetonitrile (6.11 g, 38.1 mmol) in MeOH (30 mL) is subsequently added dropwise over a period of 20 min (the addition funnel is washed with 8 mL of MeOH). The resulting mixture is then stirred at ambient temperature for 30 min. After the reaction is complete, the mixture is cooled in an ice-bath and is diluted with CH_2Cl_2 (200 mL). With vigorous stirring, 4.8 g (19 mmol) of $Na_2S_2O_3{\cdot}5\ H_2O$ and 10 g (96 mmol) of sodium carbonate are added portionwise and the resulting two-phase mixture is partitioned. The aqueous phase is extracted with CH_2Cl_2 (100 mL). The organic phase is washed with brine (100 mL) and the combined aqueous layers are back-extracted with CH_2Cl_2 (3 x 100 mL). The combined organic extracts are dried over anhydrous sodium sulfate, filtered, and concentrated to dryness on a rotary evaporator to give crude nitrone (6.4 g) as a slightly yellow solid. The nitrone is directly converted to hydroxylamine oxalate without further purification (7.51 g, 87% yield for 2 steps). The enantiomeric purity of the hydroxylamine was shown to be >99% ee upon reduction with Zn/AcOH and derivatization (see Note 21).

13. The reaction typically takes 30 min to complete. TLC analysis on Merck silica gel 60 F-254 plates eluting with MeOH:CH_2Cl_2 (2.5:97.5) shows formation of the (Z)-nitrone (R_f = 0.68) (visualized by a 254-nm UV lamp and ethanolic phosphomolybdic acid), with only a trace amount of the (E)-nitrone (R_f = 0.73).

14. Judging by ^1H NMR analysis, the crude product is highly pure (Z)-nitrone (92/8,

Z/E), which can be used for the next step. Recrystallization from EtOAc-hexane affords pure (Z)-nitrone, which exhibits the following physical properties: mp 89.5-91.0; $[\alpha]_D^{28}$ +83 (c 0.494, $CHCl_3$), $[\alpha]_D^{23}$ +135.8 (c 0.995, EtOH); IR (KBr) cm^{-1}: 3098, 2222, 1541, 1452, 1442, 1377, 1295, 1181, 1074, 1007, 748, 701; ^1H NMR (500 MHz, $CDCl_3$) δ: 1.83 (d, J = 7.0, 3 H), 5.18 (quint, J = 6.9, 1 H), 6.67 (s, 1 H), 7.43 (br s, 5 H); ^{13}C NMR (166 MHz, $CDCl_3$) δ: 19.0, 79.5, 105.8, 112.2, 128.9, 129.2, 129.8, 136.2. Anal. Calcd for $C_{10}H_{10}N_2O$: C, 68.95; H, 5.79; N, 16.08. Found: C, 68.83; H, 5.68; N, 15.90.

15. Reagent grade MeOH was purchased from Aldrich Chemical Co., Inc. and used as received.

16. Hydroxylamine hydrochloride was purchased from Aldrich Chemical Co., Inc. and used as received.

17. *Caution! The reaction should be carried out in a well-ventilated hood because of the potential of generating HCN gas.*

18. Care must be exercised not to allow the bath temperature to rise above 25°C and not to concentrate the solution to complete dryness, both of which would increase the risk of explosion of the hydroxylamine. The checkers found that some of the hydroxylamine would precipitate, resulting in a white slurry that would dissolve upon addition of the next reagent.

19. Oxalic acid was purchased from Aldrich Chemical Co., Inc. and used as received. The checkers found that sonication was helpful in dissolving the oxalic acid in MeOH.

20. The pure product exhibits the following physical properties: mp 177-180°C (dec.); $[\alpha]_D^{28}$ -2.2 (c 1.06, MeOH); IR (KBr) cm^{-1}: 3220, 2987, 2578, 1761, 1610, 1483, 1458, 1210, 986, 961, 775, 714, 702; ^1H NMR (500 MHz, CD_3OD) δ: 1.68 (d, J = 6.8, 3 H), 4.52 (q, J = 6.9, 1 H), 7.39-7.50 (m, 5 H); ^{13}C NMR (166 MHz, CD_3OD) δ: 16.1, 62.9, 129.9, 130.1, 130.6, 136.0, 166.4. Anal. Calcd for $C_8H_{11}NO{\bullet}(COOH)_2$: C, 52.86; H, 5.77; N, 6.16. Found: C, 52.72; H, 5.68; N, 6.18.

21. The enantiomeric excess of the product was determined by reduction with Zn in AcOH, followed by derivatization with 3,5-dinitrobenzoyl chloride. (S)-N-1-Phenylethylhydroxylamine oxalate (0.50 g, 2.20 mmol) was placed in a 50-mL, single-necked, round-bottomed flask equipped with an argon inlet, rubber septum, and Teflon-coated stirbar, followed by acetic acid (AcOH) (glacial, 6 mL), HCl (1M, 12 mL), and Zn dust (5 g). The slurry was heated to 80°C for 6 hr, during which time the Zn began to conglomerate. The reaction mixture was cooled to room temperature and filtered through glass wool. The Zn metal was washed with H_2O (10 mL) and CH_2Cl_2 (20 mL) and the filtrate was concentrated under reduced pressure. The residue was partitioned between CH_2Cl_2 (40 mL) and 1M NaOH (40 mL). The organic phase was washed with H_2O (20 mL) and brine (20 mL). The combined aqueous phases were back-extracted with CH_2Cl_2 (2 x 20 mL). The organic extracts were combined, dried over magnesium sulfate, and concentrated to give crude amine. Bulb-to-bulb distillation of the amine [150°C (air bath temp), 100 mm] provided (S)-1-phenylethylamine (0.21 g, 78%). In a 25-mL, two-necked, round-bottomed flask, with an argon inlet, septum, and Teflon-coated stirbar, was placed (S)-1-phenylethylamine (0.18 g, 1.5 mmol) and THF (6 mL). The solution was cooled in an ice bath to 0°C; triethylamine (0.3 mL, 2.2 mmol) and subsequently 3,5-dinitrobenzoyl chloride (0.38 g, 1.7 mmol) were added. The ice bath was removed and the solution was stirred for 2.5 hr at ambient temperature, during which time the solution turned yellow and a white precipitate formed. The slurry was then partitioned between H_2O (15 mL) and Et_2O (15 mL). The organic phase was separated, then washed with brine (10 mL). The combined aqueous phases were back-extracted with Et_2O (10 mL). The combined organic extracts were dried over magnesium sulfate, filtered, and concentrated under reduced pressure. The crude amide was purified by column chromatography (30 mm x 150 mm SiO_2, hexane/EtOAc, 4/1→2/1) to provide (S)-N-(3,5-dinitrobenzoyl)-1-phenylethylamine (0.47 g, 99%). The enantiomeric purity of the derivative was shown to be >99% ee by chiral stationary

phase HPLC and SFC. (HPLC: Pirkle S-N1N-Naphthylleucine column, hexane/i-PrOH, 3/21 mL/min, t_R 16.54 min (R isomer: t_R = 14.22 min), SFC: Chiralcel OD column, 20% MeOH, 2 mL/min, t_R = 9.136 (R isomer: 8.539 min). The submitters used reverse phase CSP HPLC for this analysis (see Note 1).

Waste Disposal Information

All toxic materials were disposed of in accordance with "Prudent Practices in the Laboratory"; National Academy Press; Washington, DC, 1995.

3. Discussion

Although N-monoalkylhydroxylamines are important precursors for N-hydroxyl-α-amino acid and hydroxamic acids,[5] there are few practical and general methods for the preparation of this class of compounds.[6] Direct oxidation of primary amines with hydrogen peroxide or MCPBA is generally ineffective, often leading to over-oxidation products. A widely used method for the conversion of primary amines to hydroxylamines employs oxidation of the corresponding Schiff bases, although the intermediacy of relatively acid-labile Schiff bases is one of its disadvantages.[7]

The present procedure provides a novel protocol for transformation of primary amines to N-monoalkylhydroxylamines in a three-step sequence involving selective mono-cyanomethylation, regioselective nitrone formation using MCPBA, and hydroxylaminolysis.[8] The procedure in Step A is representative of selective mono-cyanomethylation. Depending upon the steric bulk of the alkyl substituents, three conditions can be employed to achieve selective mono-cyanomethylation (Table 1). Since cyanomethylated amines are relatively stable, they can be used as reliable precursors for hydroxylamines. The procedure in Step B illustrates the regioselective formation of nitrone. The significance of the cyano group in directing the formation of nitrones was also demonstrated by the successful application of this method to a wide variety of substrates including α-aminoester derivatives[7b,7d,9] (Table 1). For the conversion of nitrones to hydroxylamines (Step C), we have modified the literature procedure.[7] The use of excess hydroxylamine hydrochloride at higher temperature reduced the reaction time and improved the yields. One of the advantages of the present procedure is that the oxime by-product generated after hydroxylaminolysis, namely NC-CH=NOH, can be easily removed by aqueous extraction.

1. Graduate School of Pharmaceutical Sciences, The University of Tokyo, 7-3-1 Hongo, Bunkyo-ku, Tokyo 113-0033, Japan.
2. Okawara, T.; Harada, K. *J. Org. Chem.* **1972**, *37*, 3286.
3. (a) Murahashi, S.-I.; Mitsui, H.; Shiota, T.; Tsuda, T.; Watanabe, S. *J. Org. Chem.*, **1990**, *55*, 1736-1744. (b) Murahashi, S.-I.; Shiota, T.; Imada, Y. *Org, Synth. Coll. Vol. IX*, **1998**, 632-636.
4. For a review on magnesium monoperoxyphthalate, see: Heaney, H. *Aldrichimica Acta*, **1993**, *26*, 35.
5. (a) Ottenheijm, H. C. J.; Herscheid, J. D. M. *Chem. Rev.* **1986**, *86*, 697. (b) Akiyama, A. *Yuki Gosei Kagaku* **1982**, *40*, 1189.
6. For synthesis of hydroxylamines, see: Bowman, W. R.; Marmon, R. J. In *Comprehensive Organic Functional Group Transformations*; Katritzky, A. R., FRS; Meth-Cohn, O.; Rees, C. W., FRS, Eds., Pergamon, Elsevier, Oxford, 1995; Vol. 2, Chapter 2.06 and references cited therein.
7. (a) Emmons, W. D. *J. Am. Chem. Soc.* **1957**, *79*, 5739. (b) Polonski, T.; Chimiak, A. *Tetrahedron Lett.* **1974**, 2453. (c) Wovkulich, P. M.; Uskokovic, M. R. *Tetrahedron* **1985**, *41*, 3455. (d) Grundke, G.; Keese, W.; Rimpler, M. *Synthesis* **1987**, 1115.
8. Tokuyama, H.; Kuboyama, T.; Amano, A.; Yamashita, T.; Fukuyama, T.

Synthesis, **2000**, 1299.

9. Wittman, M. D.; Halcomb, R. L.; Danishefsky, S. J. *J. Org. Chem.* **1990**, 55, 1981.

Appendix

Chemical Abstracts Nomenclature (Collective Index Number);

(Registry Number)

(S)-1-Phenylethylamine: Benzenemethanamine, α-methyl-, (αS)- (9); (2627-86-3)

Diisopropylethylamine: 2-Propanamine, N-ethyl-N-(1-methylethyl)- (9); (7087-68-5)

Bromoacetonitrile; Acetonitrile, bromo- (8, 9); (590-17-0)

(S)-[(1-Phenylethyl)amino]acetonitrile; Acetonitrile, [[(1S)-1-phenylethyl]amino]-, (9); (35341-76-5)

m-Chloroperbenzoic acid: Benzenecarboperoxoic acid, 3-chloro- (9); (937-14-4)

[(1S)-1-Phenylethyl]imino]acetonitrile N-oxide; Acetonitrile, [oxido[(1S)-1-phenylethyl]-imino]-, (9); (300843-73-6)

Hydroxylamine hydrochloride: Hydroxylamine, hydrochloride (8,9); (5470-11-1)

Oxalic acid: Ethanedioic acid (9); (144-62-7)

N-Hydroxy-(S)-1-phenylethylamine oxalate; Benzenemethanamine, N-hydroxy-α-methyl-, (αS)-, ethanedioate (1:1) salt, (9); (78798-33-1)

Table 1

Entry	Amine	Cyanomethylation[a]			Overall Yield (%) of Hydroxylamine from Cyanomethylated Amine
		Conditions	Time (h)	Yield (%)	
1	$PhCH_2NH_2$	A	24	95	82
2	$PhCH_2CH_2NH_2$	A	24	96	75
3	$PhCH_2CH_2CH_2NH_2$	A	24	93	74
4	$-NH_2$	B[b,c]	22	97	79
5	Ph_2CHNH_2	B	15	98	55
6	NH_2	C	1	89	76
7	$BnO_2CCH_2NH_2$	B[d]	18	91	62
8	CO_2Me Ph NH_2	B[b,e]	26	92	61

[a]Conditions A. $ClCH_2CN$ (1.5 eq), K_2CO_3 (2.0 eq), CH_3CN, 60°C; Conditions B. $BrCH_2CN$ (1.5 eq), i-Pr_2NEt (2.0 eq), CH_3CN, rt; Conditions C. ICH_2CN (2.0 eq), K_2CO_3 (2.5 eq), DMF, rt. [b]HCl salt of amine was used. [c]$BrCH_2CN$ (1.3 eq), i-Pr_2NEt (3.0 eq). [d]$BrCH_2CN$ (1.2 eq), i-Pr_2NEt (2.0 eq). [e]$BrCH_2CN$ (2.0 eq), i-Pr_2NEt (3.0 eq).

TETRABENZYL PYROPHOSPHATE

[Diphosphoric acid, tetrakis(phenylmethyl) ester]

$$PhCH_2O-\overset{O}{\overset{\|}{P}}(OCH_2Ph)-OH \xrightarrow[\text{Heptane, 5°C}]{\text{i-PrOAc, DCC}} PhCH_2O-\overset{O}{\overset{\|}{P}}(OCH_2Ph)-O-\overset{O}{\overset{\|}{P}}(OCH_2Ph)-OCH_2Ph$$

DBP → **TBPP**

Submitted by Todd D. Nelson,[1] Jonathan D. Rosen,[1] M. Bhupathy,[1] James McNamara,[1] Michael J. Sowa,[2] Chad Rush,[2] and Louis S. Crocker.[3]

Checked by Stuart J. Conway and Andrew B. Holmes.

1. Procedure

A flame-dried, 250-mL, three-necked, round-bottomed flask is equipped with a magnetic stir bar, low-temperature thermometer, and two rubber septa (Note 1). A balloon filled with argon is connected via an inlet needle through the central rubber septum. The vessel is charged with 15.24 g (54.8 mmol) of dibenzyl phosphate (DBP) (Notes 2, 3) and flushed with argon from the balloon, then 60 mL of isopropyl acetate is introduced through the side septum by syringe. The resulting slurry is stirred and cooled to 3 ± 3°C, and the balloon is temporarily removed from the central inlet needle while 26 mL of a 1.08M solution (28.2 mmol) of 1,3-dicyclohexylcarbodiimide (DCC) in isopropyl acetate (Notes 4, 5) *(Caution!)* is added via cannula from a flame-dried, 100-mL, round-bottomed flask equipped with another argon-inflated balloon. The reaction temperature is maintained at 3 ± 3°C by controlling the addition through the cannula by simply removing and re-inserting the needle in the flask containing the DCC (Notes 6, 7). The cold slurry is filtered through a sintered glass filter funnel under aspirator vacuum and the 1,3-dicyclohexylurea (DCU) waste cake is rinsed with 4 x 12-mL portions of isopropyl acetate. The filtrate and rinses are combined and concentrated at 25°C in vacuo on a rotary evaporator to a final volume of 30 mL (Notes 8, 9).

The concentrated solution is transferred to a 250-mL, three-necked, round-bottomed flask equipped with a magnetic stir bar, low-temperature thermometer and two rubber septa (Note 1). A balloon filled with argon is connected via an inlet needle through the central rubber septum. The solution is diluted by adding 10 mL of heptane through the septum using a syringe (Note 10). The resulting slurry is stirred at room temperature for 15 min, then 80 mL of heptane is added via syringe to the stirred slurry over 10 min (Note 11). The slurry is then cooled to 3 ± 3°C and stirred for 1 hr. The solids are collected using a sintered glass filter funnel with aspirator vacuum and the filter cake is washed with 3 x 20-mL portions of 20% (v/v) isopropyl acetate/heptane. The filter cake is dried under reduced pressure, then transferred to a 250-mL, round-bottomed flask and flushed with a blanket of argon four times at room temperature. The product was allowed to stand under an atmosphere of argon at room temperature overnight (Note 12) to afford 13.0-14.15 g (88-96%) of tetrabenzyl pyrophosphate as a white crystalline solid (Note 13).

2. Notes

1. The submitters used a 12-L, round-bottomed flask equipped with an overhead stirrer, thermocouple, N_2 inlet, and addition funnel to process 762 g of DBP.

2. The submitters carried out this preparation on fifty times the scale reported here.

3. Dibenzyl phosphate was purchased from Aldrich Chemical Company, Inc. The submitters purchased dibenzyl phosphate from Digital Specialty Company, Inc.

4. The checkers dissolved 5.8 g of DCC in 26 mL of isopropyl acetate (both purchased from Aldrich Chemical Company, Inc.). The submitters obtained a prepared solution of DCC (1.08M in isopropyl acetate) from Schweizerhall, Inc.

5. DCC, a strong sensitizer, is readily absorbed through the skin and may cause an allergic skin reaction. This compound is extremely destructive to tissue of the mucous membranes, upper respiratory tract, eyes and skin.

6. Although the dibenzyl phosphate is not completely soluble in isopropyl acetate, it is apparently sufficiently soluble to allow formation of tetrabenzyl pyrophosphate which remains in solution while a slurry of 1,3-dicyclohexylurea (DCU) forms as a white precipitate.

7. Typical addition times are between 10-20 min and the reaction is complete within 90 min (HPLC). The submitters found that addition times of 25-35 min (with 762 g of dibenzyl phosphate) were typical and the reaction was complete within 30 min.

The checkers used the following HPLC conditions:

Column:	Dynamax C18, 5 x 250 mm, 8 μm particle size
Guard Tube:	Dynamax C18 Guard, 8 μm particle size
Mobile phase:	Acetonitrile:water 70:30 (v/v)
Run time:	20 min
Flow rate:	1.5 mL/min, at room temperature

Detection at 220 nm and 260 nm (diode array detector)

Retention time for tetrabenzyl pyrophosphate = 12.0 min

The checkers found it advantageous to monitor the reaction by detection at 260 nm as well as 220 nm because at the longer wavelength there was no absorption peak due to isopropyl acetate. The final purity of the sample and the absence of isopropyl acetate were then verified by repeating the HPLC analysis using detection at 220 nm.

The submitters used the following HPLC conditions:

Column:	Zorbax RX-C8, 4.6 x 250 mm, 5 μm particle size
Mobile phase:	A) Acetonitrile
	B) 0.1% H_3PO_4 buffered water

Step gradient: 40:60 A:B to 70:30 A:B over 10 min, hold for 5 min, 70:30 A:B to 40:60 A:B over 2 min, hold for 3 min.

Flow rate: 1.5 mL/min, at room temperature

Detection at 220 nm

Retention time for tetrabenzyl pyrophosphate = 12.8 min

8. Concentration of the solution on the rotary evaporator must be ended before the solution becomes saturated. The submitters carried out this process by reduced pressure distillation. The solubility of tetrabenzyl pyrophosphate in isopropyl acetate at 25°C is ca. 676 mg/g.

9. All operating temperatures should be kept at room temperature or below due to the very large exothermic potential exhibited by dibenzyl phosphate and tetrabenzyl pyrophosphate.

10. The submitters seeded the batch with 1 mol% of tetrabenzyl pyrophosphate. The checkers found it unnecessary to use seed crystals.

11. The solvent composition of the system is approximately 4:1 heptane:isopropyl acetate. This was predetermined by using solubility data of tetrabenzyl pyrophosphate under different conditions (listed below) at 5°C.

Heptane: isopropyl acetate (v/v)	Solubility (mg TBPP/g solvent)[a]
1:4	82
2:3	34
3:2	10.6
4:1	2.8

[a] Measured at 5°C

12. The submitters dried the product cake in vacuo and under a blanket of argon overnight at room temperature.

13. The submitters obtained 671 g (91%) from 762 g of dibenzyl phosphate. Characterization data for tetrabenzyl pyrophosphate include the following: TLC on Merck

kieselgel 60 F_{254}, R_f = 0.42 (1:1 EtOAc:hexane); mp 59-60°C (lit.[4] 61-62°C, from cyclohexane); ^{1}H NMR (250 MHz, $CDCl_3$) δ: 5.10-5.13 (m, 8 H), 7.33 (s, 20 H); ^{13}C NMR (100 MHz, $CDCl_3$) δ: 70.3 (t, 4 C, J = 3), 127.9 (s, 8 C), 128.4 (s, 8 C), 128.6 (s, 4 C), 134.8 (t, 4 C, J = 4); ^{31}P NMR (162 MHz, $CDCl_3$) δ: -12.3 (s, 2 P); IR ($CHCl_3$) cm^{-1}: 3012, 1498, 1456, 1381, 1292, 1021, 956. MS (ES^+) m/z (rel intensity) 561 (100, M + Na)$^+$, 539 (50, M + H)$^+$; m/z 539.1390 ([M + H]$^+$, calcd for $C_{28}H_{29}O_7P_2$ 539.1389. Anal. Calcd for $C_{28}H_{28}O_7P_2$: C, 62.5; H, 5.2; Found: C, 62.4; H, 5.2.

Waste Disposal Information

All toxic materials were disposed of in accordance with "Prudent Practices in the Laboratory"; National Academy Press; Washington, DC, 1995.

3. Discussion

Recently, tetrabenzyl pyrophosphate (TBPP) has become increasingly utilized for either O- or N-phosphorylations in complex molecules. A wide variety of substrates have been successfully phosphorylated by the title reagent and this transformation has often been cited as a key step in many syntheses. Debenzylation of the resulting dibenzyl phosphoryl moiety is usually facile and cleanly affords the corresponding phosphate.

R₂NH or ROH —[base, (BnO)₂P(O)–O–P(O)(OBn)₂]→ —[H₂, Pd/C]→ R₂N–P(O)(OH)₂ or RO–P(O)(OH)₂

In recent years, tetrabenzyl pyrophosphate has been used to synthesize a wide variety of phosphate-containing biologically active molecules,[5] which include prodrugs of tachykinin receptor antagonists,[6] angiotensin II antagonists,[7] 5-phosphatase inhibitors,[8] DHQ synthase inhibitors,[9] EPSP synthase inhibitors,[10] HIV protease inhibitors,[11] SH2 domain inhibitors,[12] antifungal agents[13] and antitumor agents.[14]

With the increasing use of tetrabenzyl pyrophosphate,[15] especially in the pharmaceutical industry, a cost-effective and convenient multi-kilogram preparation of tetrabenzyl pyrophosphate was desirable. The current procedure is a modification of an earlier synthesis by Todd.[4] Tetrabenzyl pyrophosphate has also been synthesized from: 1) N-hydroxysuccinimide and N-mercaptosuccinimide esters[16] and 2) the corresponding silver phosphate in carbon disulfide.[17]

Tetrabenzyl pyrophosphate should be stored in a freezer as a dry solid. Stability data on TBPP are tabulated below [reported in Liquid Chromatography Area Percent (LCAP)]. The only impurity detected was dibenzyl phosphate. For this study, solid TBPP was sealed in scintillation vials with Parafilm, which were in turn enclosed in desiccators with indicating Drierite.

Table. Stability Data for TBPP

Temperature	0 days	19 days	31 days	62 days
25°C	99.9	97.1	96.9	95.5
0°C	99.9	99.5	99.0	98.4
-25°C	99.9	99.7	99.7	99.3

1. Merck Research Laboratories, Department of Process Research, P.O. Box 2000, Rahway, NJ 07065.
2. Merck Research Laboratories, Department of Chemical Engineering Research and Development, P.O. Box 2000, Rahway, NJ 07065.

3. Merck Research Laboratories, Department of Analytical Research, P.O. Box 2000, Rahway, NJ 07065.
4. Khorana, H. G.; Todd, A. R. *J. Chem. Soc.* **1953**, 2257.
5. (a) Nicolaou, M. G.; Yuan, C.-S.; Borchardt, R. T. *J. Org. Chem.* **1996**, *61*, 8636; (b) Vacca, J. P.; de Solms, S. J.; Young, S. D.; Huff, J. R.; Billington, D. C.; Baker, R.; Kulagowski, J. J.; Mawer, I. M. ACS Symposium Series 463; 1991; Chapter 5, p. 66.
6. Dorn, C. P.; Hale, J. J.; Maccoss, M.; Mills, S. G. U.S. Patent 5 691 336, 1997.
7. De Laszlo, S. E.; Glinka, T. W.; Greenlee, W. J.; Chakravarty, P. K.; Patchett, A. A. U.S. Patent 5 385 894, 1995.
8. Kozikowski, A. P.; Fauq, A. H.; Wilcox, R. A.; Nahorski, S. R. *J. Org. Chem.* **1994**, *59*, 2279.
9. Montchamp, J.-L.; Peng, J.; Frost, J. W. *J. Org. Chem.* **1994**, *59*, 6999.
10. Marzabadi, M. R.; Font, J. L.; Gruys, K. J.; Pansegrau, P. D.; Sikorski, J. A. *Bioorg. Med. Chem. Lett.* **1992**, *2*, 1435.
11. Dressman, B. A.; Fritz, J. E.; Hammond, M.; Hornback, W. J.; Kaldor, S. W.; Kalish, V. J.; Munroe, J. E.; Reich, S. H.; Tatlock, J. H.; Shepherd, T. A.; Rodriguez, M. J. U.S. Patent 5 484 926, 1996.
12. Pacofsky, G. J.; Lackey, K.; Alligood, K. J.; Berman, J.; Charifson, P. S.; Crosby, R. M.; Dorsey, Jr., G. F.; Feldman, P. F.; Gilmer, T. M.; Hummel, C. W.; Jordan, S. R.; Mohr, C.; Shewchuk, L. M.; Sternbach, D. D.; Rodriguez, M. *J. Med. Chem.* **1998**, *41*, 1894.
13. (a) Balkovec, J. M.; Black, R. M.; Hammond, M. L.; Heck, J. V.; Zambias, R. A.; Abruzzo, G.; Bartizal, K.; Kropp, H.; Trainor, C.; Schwartz, R. E.; McFadden, D. C.; Nollstadt, K. H.; Pittarelli, L. A.; Powles, M. A.; Schmatz, D. M. *J. Med. Chem.* **1992**, *35*, 194; (b) Balkovec, J. M.; Loewe, M. F.; Mathre, D. J. Eur. Patent 525 889, 1991.

14. Ueda, Y.; Mikkilineni, A. B.; Knipe, J. O.; Rose, W. C.; Casazza, A. M.; Vyas, D. M. *Bioorg. Med. Chem. Lett.* **1993**, *3*, 1761.

15. A *Chemical Abstracts* search regarding tetrabenzyl pyrophosphate (on February 1, 2002) revealed a significant increase in the use of this reagent, as indicated by the following data:

Period	Number of Citations
2002-1991	73
1990-1981	10
1980-1971	1

16. Chapman, T. M.; Kleid, D. G. *J. Org. Chem.* **1973**, *38*, 250.

17. Atkinson, R. E.; Cadogan, J. I. G. *J. Chem. Soc. (C)* **1967**, 1356.

Appendix

Chemical Abstracts Nomenclature (Collective Index Number); (Registry Number)

Tetrabenzyl pyrophosphate: Diphosphoric acid, tetrakis(phenylmethyl) ester (9); (990-91-0)

Dibenzyl phosphate (8): Phosphoric acid, bis(phenylmethyl) ester (9); (1623-08-1)

Isopropyl acetate (8): Acetic acid, 1-methylethyl ester (9); (108-21-4)

1,3-Dicyclohexylcarbodiimide (8): Cyclohexanamine, N,N'-methanetetraylbis- (9); (538-75-0)

1,3-Dicyclohexylurea: Urea, dicyclohexyl- (9); (2387-23-7)

PREPARATION OF 9,10-DIMETHOXYPHENANTHRENE AND 3,6-DIACETYL-9,10-DIMETHOXYPHENANTHRENE

(Phenanthrene, 9,10-dimethoxy- and Ethanone, 1,1'-(9,10-dimethoxy-3,6-phenanthrenediyl)bis-)

A. [9,10-phenanthrenequinone] $\xrightarrow[Me_2SO_4, H_2O, THF]{Na_2S_2O_4, KOH, Bu_4NBr}$ [9,10-dimethoxyphenanthrene]

B. [9,10-dimethoxyphenanthrene] $\xrightarrow[CH_2Cl_2]{AcCl, AlCl_3}$ [3,6-diacetyl-9,10-dimethoxyphenanthrene]

Submitted by Kamil Paruch, Libor Vyklický, and Thomas J. Katz.[1]

Checked by Mitsuru Kitamura and Koichi Narasaka.

1. Procedure

Caution: Dimethyl sulfate is highly toxic and a potential carcinogen. It should be handled with appropriate safeguards in a well-ventilated fume hood.

A. *9,10-Dimethoxyphenanthrene.* A mixture of 26 g (0.125 mol) of 9,10-phenanthrenequinone (Note 1), 13 g (0.04 mol) of tetrabutylammonium bromide (Bu_4NBr), 65 g (0.37 mol) of sodium dithionite ($Na_2S_2O_4$), 250 mL of tetrahydrofuran (THF), and 250 mL of water is shaken for 5 min in a 2-L separatory funnel. Dimethyl sulfate (62 mL, 0.65 mol) is added, followed by an aqueous solution of sodium hydroxide (64 g, 1.6 mol, in 125 mL of water) and 200 g of ice. The mixture is shaken for 5 min and, after another 200 g of ice has been added, shaken for another 10 min. Ethyl acetate (EtOAc, 100 mL) is added and the mixture is shaken. The aqueous phase is separated and extracted with EtOAc (2 x 100 mL). The combined organic extracts are washed with water (3 x 100 mL), 15% aqueous ammonia (2 x 100 mL), water (3 x 100 mL), and finally brine (100 mL). The solution is dried over sodium sulfate (Na_2SO_4) and filtered. The solvents are removed under vacuum (initially using a water aspirator, then at 0.5 torr). The residue, a thick brown oil. is dissolved in 80 mL of a ca. 2:1 mixture of dichloromethane (CH_2Cl_2) and hexane and poured onto a plug of neutral alumina, from which it is eluted with a 170-mL portion of the solvent mixture. After the solvent has been removed, the residue is dried under vacuum (ca. 0.5 torr) to give 23.7 g (80%) of 9,10-dimethoxyphenanthrene as a yellow oil (Note 2).

B. *3,6-Diacetyl-9,10-dimethoxyphenanthrene.* A 1-L, three-necked, round-bottomed flask, fitted with an addition funnel, mechanical stirrer, and hydrogen chloride trap, is charged with 23.7 g of 9,10-dimethoxyphenanthrene and 120 mL of CH_2Cl_2 (Note 3). The solution is cooled in an ice-bath, and acetyl chloride (120 mL, Note 4) is added slowly. The cooling bath is removed, and over the course of 5 min, 44 g (0.33 mol) of aluminum chloride (Note 5) is added in portions to the stirred solution. The mixture is stirred for 15 min at ambient temperature and then carefully poured onto 1 L of crushed ice. The organic phase is separated, and the aqueous phase is extracted three times with 120-mL portions of CH_2Cl_2. The combined organic phases are washed with 120 mL of water, then with 120 mL of saturated aqueous sodium bicarbonate ($NaHCO_3$). The solution is dried over Na_2SO_4 and filtered. The solvent is removed, and the residual solid is suspended in 120 mL of methanol (MeOH), filtered, washed with an additional 120 mL of MeOH, and then dried (80°C/ca. 0.5 torr) to afford 24.8 g (77% yield based on dimethoxyphenanthrene, 62% yield based on phenanthrenequinone) of 3,6-diacetyl-9,10-dimethoxyphenanthrene as a slightly pale yellow solid (Note 6).

2. Notes

1. 9,10-Phenanthrenequinone (95%), purchased from Acros, was used as received. The checkers purchased it from Tokyo Chemical Industry.

2. In two runs, the checkers obtained 27.0 and 25.7 g (91% and 90% yields). IR (neat film between NaCl plates) cm^{-1}: 2936, 1602, 1327, 1070. ^{1}H NMR ($CDCl_3$, 500 MHz in accord with the literature[2]) δ: 4.10 (s, 6H), 7.59-7.65 (m, 4H), 8.24 (dd, J = 8.0, J = 1.5, 2H), 8.64 (dd, J = 7.0, J = 1.5, 2H); ^{13}C NMR ($CDCl_3$, 75 MHz) δ: 60.9, 122.1, 122.6, 125.8, 126.8, 128.6, 129.1, 143.9. HRMS (FAB) m/z calcd for $C_{16}H_{14}O_2$: $[M]^+$ 238.0994; found 238.1003.

3. Dichloromethane was distilled from CaH_2.

4. Acetyl chloride (98%), purchased from Aldrich Chemical Company, Inc., was used as received. The checkers purchased it from Tokyo Chemical Industry.

5. Aluminum chloride (98%), purchased from Aldrich Chemical Company, Inc., was used as received. The checkers purchased it from Wako Pure Chemical Industry.

6. In two runs, the checkers obtained 26.6 and 28.8 g (82% and 83% yields). Samples can be purified further by crystallization from MeOH, but no significant amounts of impurities are detected when the material is analyzed before crystallization by means of ^{1}H NMR and ^{13}C NMR spectroscopy and HPLC (reverse phase, 5 µm LiChrospher® 100 from Hewlett-Packard, acetonitrile-H_2O gradients varying from 80:20 to 100:0). The pure product is a white solid, mp 160–161°C. IR (CCl_4) cm^{-1}: 2939, 1687, 1611, 1318, 1059. ^{1}H NMR ($CDCl_3$, 400 MHz) δ: 2.82 (s, 6 H), 4.13 (s, 6 H), 8.20 (dd, J = 8.5, 1.3, 2 H), 8.32 (d, J = 8.5, 2 H), 9.31 (d, J = 1.3, 2 H); ^{13}C NMR ($CDCl_3$, 75 MHz) δ: 27.0, 61.1, 122.8, 123.6, 126.3, 128.5, 132.6, 134.7, 145.5, 198.0. UV-vis (CH_3CN, c = 1.81 x 10^{-4} M): λ_{max} (log ε) 266 (4.05), 325 nm (4.06). Anal. Calcd for $C_{20}H_{18}O_4$: C, 74.52; H, 5.63. Found: C, 74.52; H; 5.66.

Waste Disposal Information

All toxic materials were disposed of in accordance with "Prudent Practices in the Laboratory"; National Academy Press; Washington, DC, 1995.

3. Discussion

Almost no 9,10-dialkoxyphenanthrenes are known, and the quantities of those few made by other methods are either small or unstated.[2,3] The best alternative procedure gave 9,10-dimethoxyphenanthrene when 9,10-phenanthrenequinone was combined first with sodium in diglyme and then with dimethyl sulfate, but the yield after two chromatographic purifications was 51%.[2b] The only other reported alkylations of phenanthrene-9,10-diol are by 1-chloro-2-diethylaminoethane in 10% yield,[3f] and by 1,2-di(bromomethyl)benzene in 22% yield.[3g] Unlike the procedure recorded here, none of these employ two-phase alkylations catalyzed by phase transfer agents.[4]

3,6-Diacetyl-9,10-dimethoxyphenanthrene has not been prepared by any other method. However, two related compounds, 9,10-di-TBDMSO- and 9,10-diphenylmethylenedioxyphenanthrene, have been prepared by brominating phenanthrene-9,10-dione, converting the diones to the ethers, coupling the dibromides with tributyl(1-ethoxyethenyl)stannane by the Stille method, and hydrolyzing the enol ethers.[5] The procedure described here circumvents three disadvantages of the tin reagents used in the Stille procedure: expense, toxicity, and the need for chromatography to remove side products, mainly Bu_3SnBr. 3,6-Diacetyl-9,10-dialkoxyphenanthrene, like the 9,10-di-TBDMSO- and 9,10-diphenylmethylenedioxy-derivatives, combines with 1,4-benzoquinone to give a [7]helicenebisquinone.[6]

There are no reported alternatives to the current procedure for the acylation of a phenanthrene using the Friedel-Crafts reaction.[7,8] Indeed, alternative methods to cleanly prepare 3,6-disubstituted derivatives of phenanthrene by means of electrophilic substitution are not known,[9,10,11] nor is there a precedent for the electrophilic substitution of any 9,10-dialkoxyphenanthrene.

1. Department of Chemistry, Columbia University, New York, NY 10027.
2. (a) Eichner, M.; Merz, A. *Tetrahedron Lett.* **1981**, *22*, 1315; (b) Rio, G.; Berthelot, J. *Bull. Soc. Chim. France* **1972**, 822.
3. (a) Dannenberg, H.; Keller, H.-H. *Chem. Ber.* **1967**, *100*, 23; (b) Adams, C.; Kamkar, N. M.; Utley, J. H. P. *J. Chem. Soc., Perkin Trans. 2* **1979**, 1767; (c) Sucharda-Sobczyk, A.; Syper, L. *Rocz. Chem.* **1975**, *49*, 749; (d) Santamaria, J.; Ouchabane, R. *Tetrahedron* **1986**, *42*, 5559; (e) Goldschmidt, S.; Schmidt, W. *Chem. Ber.* **1922**, *55*, 3197; (f) Fourneau, E.; Matti, J. *Bull. Soc. Chim. France* **1942**, 633; (g) Kurebayashi, H.; Mine, T.; Harada, K.; Usui, S.; Okajima, T.; Fukazawa, Y. *Tetrahedron* **1998**, *54*, 13495.
4. Dehmlow, E. V.; Dehmlow, S. S. "Phase Transfer Catalysis"; VCH: New York, 1993.
5. Fox, J. M.; Goldberg, N. R.; Katz, T. J. *J. Org. Chem.* **1998**, *63*, 7456.
6. Paruch, K.; Vyklický, L.; Katz, T. J. *Org. Synth.* **2003**, *80*, 233.
7. Monoacylations: (a) Mosettig, E.; van de Kamp, J. *J. Am. Chem. Soc.* **1930**, *52*, 3704; (b) Gore, P. H. *J. Org. Chem.* **1957**, *22*, 135; (c) Blin, P.; Bunel, C.; Maréchal, E. *J. Chem. Res. (S)* **1978**, 206; (d) Fernández, F.; Gómez, G.; López, C.; Santos, A. *J. Prakt. Chem.* **1989**, *331*, 15.
8. The only diacylations are by phthalic anhydride with aluminum chloride in "tetrachloroethane" and by benzoyl chloride with $AlCl_3$ in the absence of additional solvent. The diketones were isolated in 15% and unspecified yields, respectively, and their structures were not determined. Clar, E.; Kelly, W. *J. Am. Chem. Soc.* **1954**, *76*, 3502.
9. Upon sulfonation, phenanthrene-3-sulfonic acid gives mainly (but not exclusively) the 3,6-disulfonic acid (Fieser, L. F. *J. Am. Chem. Soc.* **1929**, *51*, 2471), but the

starting phenanthrene-3-sulfonic acid can be obtained from phenanthrene in only ca. 25% yield, and then isolated only from a mixture including a comparable amount of the 2-sulfonic acid (Fieser, L. F. *Org. Synth., Coll. Vol. II* **1943**, 482).

10. Cerfontain, H.; Koeberg-Telder, A.; Laali, K.; Lambrechts, H. J. A. *J. Org. Chem.* **1982**, *47*, 4069.
11. Schmidt, J.; Schairer, O. *Chem. Ber.* **1923**, *56*, 1331.

Appendix

Chemical Abstracts Nomenclature (Collective Index Number); (Registry Number)

9,10-Phenanthrenedione (9); (84-11-7)

9,10-Dimethoxyphenanthrene: Phenanthrene, 9,10-dimethoxy- (8, 9); (13935-65-4)

Tetrabutylammonium bromide: 1-Butanaminium, N,N,N-tributyl-, bromide (9); (1643-19-2)

Sodium dithionite: Dithionous acid, disodium salt (8CI, 9CI); (7775-14-6)

Dimethyl sulfate: Sulfuric acid, dimethyl ester (8, 9); (77-78-1)

Acetyl chloride (8, 9); (75-36-5)

Aluminum chloride: Aluminum chloride ($AlCl_3$) (9); (7446-70-0)

3,6-Diacetyl-9,10-dimethoxyphenanthrene: Ethanone, 1,1'-(9,10-dimethoxy-3,6-phenanthrenediyl)bis- (9); (310899-08-2)

HELICENEBISQUINONES: SYNTHESIS OF A [7]HELICENEBISQUINONE (Dinaphtho[2,1-c:1',2'-g]phenanthrene-1,4,15,18-tetrone, 9,10-dimethoxy-6,13-bis[[tris(1-methylethyl)silyl]oxy]-)

A. MeO, MeO (3,6-diacetyl-9,10-dimethoxyphenanthrene) —TIPSOTf, TEA, CH_2Cl_2→ 3,6-bis[1-(OTIPS)ethenyl]-9,10-dimethoxyphenanthrene (MeO, MeO, OTIPS, OTIPS)

B. 3,6-bis[1-(OTIPS)ethenyl]-9,10-dimethoxyphenanthrene (MeO, MeO, OTIPS, OTIPS) —1,4-benzoquinone, heptane, reflux→ [7]helicenebisquinone (TIPSO, OMe, OMe, OTIPS, O, O, O, O)

Submitted by Kamil Paruch, Libor Vyklický, and Thomas J. Katz.[1]

Checked by Mitsuru Kitamura and Koichi Narasaka.

1. Procedure

A. *3,6-Bis[1-(triisopropylsiloxy)ethenyl]-9,10-dimethoxyphenanthrene.* A solution of 8 g (0.025 mol) of 3,6-diacetyl-9,10-dimethoxyphenanthrene (Note 1) and 28 mL of triethylamine (Et_3N) (Note 2) in 80 mL of dry dichloromethane (CH_2Cl_2) (Note 2) is prepared under nitrogen in a 250-mL, round-bottomed flask equipped with a magnetic stir bar. The flask is cooled in an ice bath and 14 mL (0.052 mol) of triisopropylsilyl triflate (Note 3) is slowly added by syringe. The mixture is stirred for 15 min while cooling in the ice bath, then for 1 hr at ambient temperature. Hexane (200 mL) is added, and the organic phase is washed twice with 100-mL portions of 1M aqueous potassium hydroxide (KOH), dried quickly over potassium carbonate (K_2CO_3), and filtered. The solvent is removed, and the oily residue is dissolved in ca. 40 mL of hexane. The solution is poured onto a 4 x 4-cm plug of neutral alumina and washed with 200 mL of hexane. The solvent is removed and the residue is dried under vacuum (0.5 torr/80°C). 3,6-Bis[1-(triisopropylsiloxy)ethenyl]-9,10-dimethoxyphenanthrene (14.4 g, 91%) is obtained as a yellow oil (Notes 4, 5) and used in the subsequent step without further purification.

B. *6,13-Bis(triisopropylsiloxy)-9,10-dimethoxy[7]helicenebisquinone.* A solution of the material obtained in Step A in 100 mL of heptane (Notes 6,7) is added under nitrogen to a 500-mL, round-bottomed flask fitted with a reflux condenser. Thirty-four grams (0.31 mol) of 1,4-benzoquinone is added (Notes 8, 9), and the mixture is heated with an oil bath at 120°C for 3.5 days with stirring (Note 10). The reaction mixture is then cooled to 25°C and 25 mL of CH_2Cl_2 is added. The solids are broken into small pieces with the aid of a spatula, and the mixture is shaken for 15 min with 50 g of sand (Note 11). The supernatant liquid is decanted, the residue is extracted three times with 70-mL portions of 1:1 hexane:CH_2Cl_2, and the combined extracts are filtered through a pad of Celite. The solution is concentrated, and 1,4-benzoquinone is removed by sublimation by heating the residue to 100°C and applying a vacuum of ca. 0.5 torr. Methanol (MeOH, 100 mL) and ca. 20 g of sand are added, and the mixture is shaken until the solids are finely suspended (ca. 30 min). Water (20 mL) is added and the mixture is shaken for 10 min. The solids are collected by filtration and washed with ca. 250 mL of 5:1 MeOH:H_2O. The powder obtained after drying under vacuum at 100°C is dissolved in a small amount of toluene and purified by flash chromatography[2] on a 12 x 8-cm column of silica gel. A purple impurity is eluted with toluene, and the product is eluted with 1-3% tetrahydrofuran (THF) in toluene. After removal of the solvents and drying under vacuum at 100°C, 4.5 g of the helicenebisquinone (a 25% yield based on the enol ether, 22% based on 3,6-diacetyl-9,10-dimethoxyphenanthrene) is obtained as a dark red solid (Note 12).

2. Notes

1. Prepared according to the accompanying procedure (Paruch, K.; Vyklický, L.; Katz, T. J. *Org. Synth.* **2003**, *80*, 227).

2. Dichloromethane and triethylamine were distilled from CaH_2.

3. Triisopropyl triflate (97%) was purchased from GSF Chemicals and used as received. The checkers purchased it from Tokyo Chemical Industry.

4. The product solidifies occasionally.

5. In two runs, the checkers obtained 15.8 and 15.2 g (99% and 98% yields). The product exhibits the following properties: ^1H NMR ($CDCl_3$, 400 MHz) δ: 1.18 (d, 36 H, J = 7.3), 1.35 (m, 6 H), 4.11 (s, 6 H), 4.57 (d, 2 H, J = 1.8), 5.06 (d, 2 H, J = 1.8), 7.90 (dd, 2 H, J = 8.6, 1.6), 8.17 (d, 2 H, J = 8.6), 8.96 (d, 2 H, J = 1.5); ^{13}C NMR ($CDCl_3$, 75 MHz) δ: 12.8, 18.2, 61.0, 90.8, 119.5, 122.0, 124.3, 128.6, 129.1, 135.4, 144.1, 156.4; IR (CCl_4) cm^{-1}: 2946, 2868, 1607, 1464, 1322, 1292, 1113, 1015. The checkers reported the following elemental analysis. Calcd. for $C_{30}H_{58}O_4Si_2$: C, 71.87; H, 9.21. Found: C, 71.65; H, 9.20.

6. Heptane was distilled from sodium/benzophenone.

7. The yield is lower when toluene is used as solvent.

8. 1,4-Benzoquinone (98%, Aldrich) is purified by slurrying it in CH_2Cl_2 with two times its weight of basic alumina, filtering the mixture through Celite, concentrating the filtrate, and drying the residue under vacuum.

9. When seven molar equivalents of 1,4-benzoquinone are used, the yield drops to 12%.

10. Sometimes sublimation of 1,4-benzoquinone plugs the condenser, but turning off the cooling water ameliorates the problem.

11. The 20-30 mesh sand is obtained from Fisher Scientific.

12. In two runs the checkers obtained 8.4 and 8.5 g (40% and 42% yields). They found that the product can be purified further by recrystallization from MeOH/EtOAc; mp 272–274°C. IR (CCl_4) cm^{-1}: 2948, 2870, 1665, 1610, 1573, 1471, 1385, 1295, 1096. ^1H NMR ($CDCl_3$, 400 MHz) δ: 1.20 (d, 18 H, J = 7.5), 1.24 (d, 18 H, J = 7.5), 1.51 (m, 6 H), 4.21 (s, 6 H), 5.92 (d, 2 H, J = 10.1), 6.48 (d, 2 H, J = 10.1), 7.38 (s, 2 H), 8.40 (d, 2 H, J = 8.9), 8.46 (d, 2 H, J = 8.9); ^{13}C NMR ($CDCl_3$, 75 MHz) δ: 13.0, 18.1, 61.2, 107.7, 121.1, 123.3, 125.0, 126.1, 128.0, 129.6, 129.9, 133.0, 134.1, 140.4, 145.3, 157.4, 183.6, 184.6. UV-vis (CH_3CN, c = 5.50×10^{-5} M): λ_{max} nm (log ε) 241 (4.54), 285 (4.48), 342 (4.18), 417 (3.76). Anal. Calcd for $C_{50}H_{58}O_8Si_2$: C, 71.22; H, 6.93. Found: C, 70.99; H, 6.87. The checkers obtained C, 71.20; H, 6.93.

Waste Disposal Information

All toxic materials were disposed of in accordance with "Prudent Practices in the Laboratory"; National Academy Press; Washington, DC, 1995.

3. Discussion

The procedure illustrates one that has been used to prepare significant quantities of a variety of helicenebisquinones.[3] The majority of helicenes have been synthesized by photocyclization of derivatives of stilbene,[4] but in most cases only small amounts can be prepared practically. Moreover, the functional groups that have been incorporated in these helicenes have not been generally useful. Using our procedure, helicenebisquinones having functional groups suitable for subsequent transformations can be prepared easily in significant amounts. Among the structures synthesized from these helicenes are helical polymers,[5] helical discotic columnar aggregates[3e,6] and liquid crystals,[7] helical ligands,[8] nonlinear optical materials,[9] asymmetric catalysts,[10] and substances that probe remote chirality.[11]

Similar procedures have been used to prepare [7]helicenes that have different side chains,[3a,b] [5]- and [6]carbohelicenes,[3c,d,e,g] and [7]heterohelicenes.[3f, h]

1. Department of Chemistry, Columbia University, New York, NY 10027.
2. Still, W. C.; Kahn, M.; Mitra, A. *J. Org. Chem.* **1978**, *43*, 2923.
3. (a) Fox, J. M.; Goldberg, N. R.; Katz, T. J. *J. Org. Chem.* **1998**, *63*, 7456; (b) Paruch, K.; Katz, T. J.; Incarvito, C.; Lam, K.-C.; Rhatigan, B.; Rheingold, A. L., *J. Org. Chem.* **2000**, *65*, 7602; (c) Katz, T. J.; Liu, L.; Willmore, N. D.; Fox, J. M.; Rheingold, A. L.; Shi, S.; Nuckolls, C.; Rickman, B. H. *J. Am. Chem. Soc.* **1997**, *119*, 10054; (d) Dreher, S. D.; Katz, T. J.; Lam, K.-C.; Rheingold, A. L. *J. Org. Chem.*, **2000**, *65*, 815; (e) Nuckolls, C.; Katz, T. J.; Katz, G.; Collings, P. J.; Castellanos, L. *J. Am. Chem. Soc.* **1999**, *121*, 79; (f) Dreher, S. D.; Weix, D. J.; Katz, T. J. *J. Org. Chem.* **1999**, *64*, 3671; (g) Paruch, K.; Vyklický , L.; Katz, T. J.; Incarvito, C. D.; Rheingold, A. L. *J. Org. Chem.* **2000**, *65*, 8774; (h) Phillips, K. E. S.; Katz, T. J.; Jockush, S.; Lovinger, A. J.; Turro, N. J. *J. Am. Chem. Soc.* **2001**, *123*, 11899.
4. (a) Laarhoven, W. H.; Prinsen, W. J. *Top. Curr. Chem.* **1984**, *125*, 63. (b) Mallory, F. B.; Mallory, C. W. *Organic Reactions*; Wiley: New York, **1984**; Vol. 30, p. 1.
5. Dai, Y.; Katz, T. J. *J. Org. Chem.* **1997**, *62*, 1274.
6. (a) Lovinger, A. J.; Nuckolls, C.; Katz, T. J. *J. Am. Chem. Soc.* **1998**, *120*, 264; (b) Nuckolls, C.; Katz, T. J.; Verbiest, T.; Van Elshocht, S.; Kuball, H.-G.; Kiesewalter, S.; Lovinger, A. L.; Persoons, A. *J. Am. Chem. Soc.* **1998**, *120*, 8656; (c) Fox, J.

M.; Katz, T. J.; Van Elshocht, S.; Verbiest, T.; Kauranen, M.; Persoons, A.; Thongpanchang, T.; Krauss, T.; Brus, L. *J. Am. Chem. Soc.* **1999**, *121*, 3453.

7. (a) Nuckolls, C.; Katz, T. J. *J. Am. Chem. Soc.* **1998**, *120*, 9541; (b) Nukolls, C,; Shao, R.; Jang, W.-G.; Clark, N. A.; Walba, D.; Katz, T. J. *Chem. Mater.* **2000**, *14*, 773.
8. Fox, J. M.; Katz, T. J. *J. Org. Chem.* **1999**, *64*, 302.
9. (a) Verbiest, T.; Van Elshocht, S.; Kauranen, M.; Hellemans, L.; Snauwaert, J.; Nuckolls, C.; Katz, T. J.; Persoons, A. *Science (Washington, D. C.)* **1998**, *282*, 913; (b) Busson, B.; Kauranen, M.; Nuckolls, C.; Katz, T. J. Persoons, A. *Phys. Rev. Lett.*, **2000**, *84*, 79; (c) Verbiest, T.; Sioncke, S.; Persoons, A.; Vyklický , L.; Katz, T. J. *Angew. Chem. Int. Ed.* **2002**, *14*, 3882.
10. Dreher, S. D.; Katz, T. J.; Lam, K.-C.; Rheingold, A. L. *J. Org. Chem.*, **2000**, *65*, 815.
11. Weix, D. J.; Dreher, S. D.; Katz, T. J. *J. Am. Chem. Soc.* **2000**, *122*, 10027.

Appendix

Chemical Abstracts Nomenclature (Collective Index Number); (Registry Number)

3,6-Diacetyl-9,10-dimethoxyphenanthrene: Ethanone, 1,1'-(9,10-dimethoxy-3,6-phenanthrenediyl)bis- (9); (310899-08-2)

3,6-Bis[1-(triisopropylsiloxy)ethenyl]-9,10-dimethoxyphenanthrene: Silane, [(9,10-dimethoxy-3,6-phenenthrenediyl)bis(ethenylideneoxy)]tris(1-methylethyl)- (9);

Triisopropyl triflate: Methanesulfonic acid, trifluoro-, tris(1-methylethyl)silyl ester (9); (80522-42-5)

Triethylamine: Ethanamine, N,N-diethyl- (9); (121-44-8)

1,4-Benzoquinone: 2,5-Cyclohexadiene-1,4-dione (9); (106-51-4)

6,13-Bis(triisopropylsiloxy)-9,10-dimethoxy[7]helicenebisquinone :Dinaphtho [2,1-c:1',2'-g]phenanthrene-1,4,15,18-tetrone, 9,10-dimethoxy-6,13-bis[[tris(1-methylethyl)silyl]oxy]- (9); (310899-14-0)

Unchecked Procedures

Accepted for checking during the period September 1, 2001 through October 1, 2002. An asterisk(*) indicates that the procedure has been subsequently checked.

Previously, *Organic Syntheses* has supplied these procedures upon request. However, because of the potential liability associated with procedures which have not been tested, we shall continue to list such procedures but requests for them should be directed to the submitters listed.

2966 Synthesis of 9-Spiroepoxy-endo-tricyclo[5.2.2.0^{2,6}]undeca-4,10-dien-8-one.
V. Singh, M. Porinchu, P. Vedantham, and P. K. Sahu, Department of Chemistry, Indian Institute of Technology, Powai, Bombay 400 076, India

2968 (R)-(+)-3,4-Dimethylcyclohex-2-en-1-one
J. D. White and U. M. Grether, Department of Chemistry, Oregon State University, Corvallis, OR 97331-4003

2975 Synthesis of Diethyl 3-Oxobicyclo[3.3.0]0ct-1-ene-7,7-dicarboxylate
M. Hayashi, T. Hino, and K. Saigo, Department of Chemistry and Biotechnology, Graduate School of Engineering, The University of Tokyo, Hongo, Bunkyo-ku, Tokyo 113-8656, Japan

2977 Synthesis and Use of Glycosyl Phosphates as Glycosyl Donors
K. R. Love and P. H. Seeberger, Department of Chemistry, Massachusetts Institute of Technology, M.I.T. Room 18-211, Cambridge, MA 02139

2980 (2S,5S)-2-(tert-Butyl)-5-methyl-1,3-dioxan-4-one
F. Lang, I. Houpis, D. Zewge, Z. J. Song, R. P. Volante, P. J. Reider, M. Kawasaki, and S. Kato, Process Research, Merck Research Laboratories, P.O. Box 2000, RY 800-C264, Rahway, NJ 07065

2983A Palladium-Catalyzed Cross-Coupling of (Z)-1-Heptenyldimethylsilanol with 4-Iodoanisole
S. E. Denmark and Z. Wang, Department of Chemistry, 600 S. Matthews Ave., University of Illinois, Urbana, IL 61801

2983B Platinum-Catalyzed Hydrosilylation and Palladium-Catalyzed Cross-Coupling: One Pot Hydroarylation of 1-Heptyne
S. E. Denmark and Z. Wang, Department of Chemistry, 600 S. Matthews Ave., University of Illinois, Urbana, IL 61801

S. E. Denmark, J. F. Fu, and M. J. Lawler, Department of Chemistry, 600 S. Matthews Ave., University of Illinois, Urbana, IL 61801

2985 Catalytic Asymmetric Synthesis of 3,3'-BINOL Derivatives Using a 1,5-Diaza-cis-Decalin-Copper(I) Iodide Complex: Dimethyl (1R)-2,2'-Dihydroxy-1,1'-binaphthalene-3,3'-dicarboxylate

X. Zu, J. Yang, and M. C. Kozlowski, Department of Chemistry, University of Pennsylvania, Room 4002 Roy and Diana Vagelos Laboratories, Philadelphia, PA 19104-6323

2986 Synthesis and Resolution of 1,5-Diaza-cis-Decalin

X. Zu, X. Li, J. D. Flesch, C. Hoess, and M. C. Kozlowski, Department of Chemistry, University of Pennsylvania, Room 4002 Roy and Diana Vagelos Laboratories, Philadelphia, PA 19104-6323

2987 A. Facile Synthesis of Aminocyclopropanes

A. deMeijere, H. Winsel, and B. Stecker, Institut für Organische Chemie, Georg-August-Universitat Göttingen, Tammanstrasse 2, D-37077 Göttingen, Germany

2989 Heck Reactions with Aryl Chlorides Catalyzed by Palladium/Tri-tert-butylphosphine: (E)-2-Methyl-3-phenylacrylic Acid Butyl Ester and (E)-4-(2-Phenylethenyl)benzonitrile

A. F. Littke and G. D. Fu, Department of Chemistry, Massachusetts Institute of Technology, Cambridge, MA 02139

2991 Methyltrioxorhenium-Catalyzed Oxidation of Secondary Amines to Nitrones: N-Benzylidene-Benzylamine N-Oxide

A. Goti, F. Cardona, and G. Soldani, Dipartimento di Chimica Organica "Ugo Schiff", Polo Scientifico-Universita' di Firenze, via della Lastruccia 13, I-50019 Sesto Fiorentino, Firenze, Italy

2992 Preparation of (Z)-4-Iodo 4 dooon-2-one from 1-Octyne

F.-T. Luo and C. Xue, Institute of Chemistry, Academia Sinica, Taipei, Taiwan 115, Republic of China

AUTHOR INDEX FOR VOLUME 80

This index comprises the names of contributors to Volume **80**, only. For authors of previous volumes, see either indices in Volume **79,** Collective Volumes I through IX, or the single volume entitled *Organic Syntheses, Collective Volumes I-VIII, Cumulative Indices,* edited by J. P. Freeman.

SUBJECT INDEX FOR VOLUME 80

This index comprises subject matter for Volume **80** only. For subjects in previous volumes, see either the indices in Collective Volumes I through IX, Volume 79, or the single volume entitled *Organic Syntheses, Collective Volumes I-VIII, Cumulative Indices,* edited by J. P. Freeman.

The index lists the names of compounds in two forms. The first is the name used commonly in procedures. The second is the systematic name according to **Chemical Abstracts** nomenclature, accompanied by its registry number in parentheses. Also included are general terms for classes of compounds, types of reactions, special apparatus, and unfamiliar methods.

Most chemicals used in the procedure will appear in the index as written in the text. There generally will be entries for all starting materials, reagents, intermediates, important by-products, and final products. Entries in capital letters indicate compounds, reactions, or methods appearing in the title of the preparation.

Ammonia; (7664-41-7), **80**, 31

p-Anisidine: Benzenamine, 4-methoxy-; (104-94-9), **80**, 160

Benzaldehyde; (100-52-7), **80**, 160

Benzenethiol; Thiophenol; (108-98-5), **80**, 184

(4S)-2-(BENZHYDRYLIDENAMINO)PENTANEDIOIC ACID, 1-tert-BUTYL ESTER-5-METHYL ESTER: L-Glutamic acid, N-(diphenylmethylene)-, 1-(1,1-dimethylethyl) 5-methyl ester; (212121-62-5), **80**, 38

1,4-Benzoquinone: 2,5-Cyclohexadiene-1,4-dione; (106-51-4), **80**, 233

N-Benzylidene-p-anisidine: Benzenamine, 4-methoxy-N-(phenylmethylene)-; (783-08-4), **80**, 160

trans-2-BENZYL-1-METHYLCYCLOPROPAN-1-OL; 1-Cyclopropanol, 1-methyl-2-phenylmethyl-; **80**, 111

BICYCLO[3.1.0]HEXAN-1-OL; (7422-09-5), **80**, 111

2,2-Bis(chloromethyl)-5,5-dimethyl-1,3-dioxane: 1,3-Dioxane, 2,2-bis(chloromethyl)-5,5-dimethyl-; (133961-12-3), **80**, 144

Bis(tricyclohexylphosphine)benzylidine ruthenium(IV) dichloride: Ruthenium, dichloro(phenylmethylene)bis(tricyclohexylphosphine)-; (172222-30-9), **80**, 85

6,13-BIS(TRIISOPROPYLSILOXY)-9,10-DIMETHOXY[7]HELICENEBISQUINONE: Dinaphtho[2,1-c:1',2'-g]phenanthrene-1,4,15,18-tetrone, 9,10-dimethoxy-6,13-bis[[tris(1-methylethyl)silyl]oxy]-; (310899-14-0), **80**, 233

3,6-Bis[1-(triisopropylsiloxy)ethenyl]-9,10-dimethoxyphenanthrene: Silane, [(9,10-dimethoxy-3,6-phenenthrenediyl)bis(ethenylideneoxy)]tris(1-methylethyl)-; **80**, 233

Boc-diallylamine: Carbamic acid, di-2-propenyl-, 1,1-dimethylethyl ester; (151259-38-0), **80**, 85

N-Boc-3-PYRROLINE: 1H-Pyrrole-1-carboxylic acid, 2,5-dihydro-, 1,1-dimethylethyl ester; (73286-70-1), **80**, 85 endo-1-BORNYLOXYETHYL ACETATE: Ethanol, 1-[[(1R,2S,4R)-1,7,7-trimethylbicyclo[2.2.1]-hept-2-yl]oxy]-, acetate; (284036-61-9), **80**, 177

Bromine; (7726-95-6), **80**, 75

Bromoacetonitrile; Acetonitrile, bromo-; (590-17-0), **80**, 207

Bromobenzene: Benzene, bromo-; (100-86-1), **80**, 57

1-Bromo-2-propyne: 1-propyne, 3-bromo-; (106-96-7), **80**, 93

(R)-(N-tert-BUTOXYCARBONYL)ALLYLGLYCINE: 4-Pentenoic acid, 2-[[(1,1-dimethylethoxy)carbonyl]amino]-, (2R)-; (170899-08-8), **80**, 31

(3R,5R,6S)-4-tert-Butoxycarbonyl-5,6-diphenyl-3-(1'-prop-2'-enyl)morpholin-2-one: 4-Morpholinecarboxylic acid, 2-oxo-5,6-diphenyl-3-(2-propenyl)-, 1,1-dimethylethyl ester, [3R-(3α,5β,6β)]-; (143140-32-3), **80**, 31

(S)-N-(tert-Butoxycarbonyl)valine methyl ester: Valine, N-[(1,1-dimethylethoxy)-carbonyl]-, methyl ester, (S)-; (58561-04-9), **80**, 57

(S)-N-(tert-Butoxycarbonyl)valine: Valine, N-[(1,1-dimethylethoxy)carbonyl]-; (13734-41-3), **80**, 57

n-Butyl acrylate: 2-Propenoic acid, butyl ester; (141-32-2), **80**, 172

n-BUTYL 2,2-DIFLUOROCYCLOPROPANECARBOXYLATE: Cyclopropanecarboxylic acid, 2,2-difluoro-, butyl ester; (260352-79-2), **80**, 172

Butyllithium: Lithium, butyl-; (109-72-8), **80**, 160

sec-Butyllithium: Lithium, (1-methylpropyl)-; (598-30-1), **80**, 46

n-Butylmagnesium chloride: Magnesium, butylchloro-; (693-04-9), 111

(5S,6R)-4-tert-BUTYLOXYCARBONYL-5,6-DIPHENYLMORPHOLIN-2-ONE: 4-Morpholinecarboxylic acid, 6-oxo-2,3-diphenyl-, 1,1-dimethylethyl ester, (2R-cis)-; (173397-90-5), **80**, 18, 31

(5R,6S)-4-tert-BUTYLOXYCARBONYL-5,6-DIPHENYLMORPHOLIN-2-ONE: 4-Morpholinecarboxylic acid, 6-oxo-2,3-diphenyl-, 1,1-dimethylethyl ester, (2S-cis)-; (112741-50-1), **80**, 18

D-(+)-Camphorsulfonic acid: Bicyclo[2.2.1]heptane-1-methanesulfonic acid, 7,7-dimethyl-2-oxo-, (1S,4R)-; (3144-16-9), **80**, 66

(2-Carbomethoxy-6-nitrobenzyl)triphenylphosphonium bromide: Phosphonium, [[2-(methoxycarbonyl)-6-nitrophenyl]methyl]triphenyl-, bromide; (195992-09-7), **80**, 75

Carbon dioxide; (124-38-9), **80**, 46

Carbon monoxide; (630-08-0), **80**, 75, 93

Carbonyldihydridotris(triphenylphosphine)ruthenium(II): Ruthenium, carbonyldihydridotris(triphenylphosphine); (25360-32-1), **80**, 104

Cesium hydroxide: Cesium hydroxide (CsOH); (21351-79-1), **80**, 38

Chloroacetyl chloride: Acetyl chloride, chloro-; (79-04-9), **80**, 200

2-CHLORO-1,3-BIS(DIMETHYLAMINO)TRIMETHINIUM HEXAFLUOROPHOSPHATE: Methanaminium, N-[2-chloro-3-(dimethylamino)- 2-propenylidene]-N-methyl-, hexafluorophosphate(1-); (249561-98-6), **80**, 207

m-Chloroperbenzoic acid; Benzenecarboperoxoic acid, 3-chloro-; (937-14-4), **80**, 207

9-Chloromethylanthracene: Anthracene, 9-(chloromethyl)-; (24463-19-2), **80**, 38

N-Chlorosuccinimide: 2,5-Pyrrolidinedione, 1-chloro-; (128-09-6), **80**, 133

Chlorotitanium triisopropoxide: Titanium, chlorotris(2-propanolato)-, (T-4)-; (20717-86-6), **80**, 111

Chlorotrimethylsilane: Silane, chlorotrimethyl-; (75-77-4), **80**, 172

Cinchonidine: Cinchonan-9-ol, (8α, 9R)-; (485-71-2), **80**, 38

R-(+)-Citronellal: 6-Octenal, 3,7-dimethyl-, (3R)-; (2385-77-5), **80**, 195

Copper(I) Iodide: Copper iodide (CuI); (7681-65-4), **80**, 129

Cyclohexylamine: Cyclohexanamine; (108-91-8), **80**, 93

2-Cyclopenten-1-one; (930-30-3), **80**, 144

Cyclopentylmagnesium chloride: Magnesium, chlorocyclopentyl-; (32916-51-1), **80**, 111

Deuterium oxide: Water-d2; (7789-20-0), **80**, 120

3,6-DIACETYL-9,10-DIMETHOXYPHENANTHRENE: Ethanone, 1,1'-(9,10-dimethoxy-3,6-phenanthrenediyl)bis-; (310899-08-2), **80**, 227, 233

Dibenzofuran; (132-64-9), **80**, 46

Dibenzofuran-4,6-dicarbonyl chloride: 4,6-Dibenzofurandicarbonyl dichloride; (151412-73-8), **80**, 46

Dibenzofuran-4,6-dicarboxylic acid: 4,6-Dibenzofurandicarboxylic acid; (88818-47-7), **80**, 46

(R,R)-Dibenzofuran-4,6-dicarboxylic acid bis(2-hydroxy-1-phenylethyl)amide: 4,6-Dibenzofurandicarboxamide, N,N'-bis[(1R)-2-hydroxy-1-phenylethyl]-; (247097-79-6), **80**, 46

(R,R)-4,6-DIBENZOFURANDIYL-2,2'-BIS(4-PHENYLOXAZOLINE) (DBFOX/PH): Oxazole, 2,2'-(4,6-dibenzofurandiyl)bis(4,5-dihydro-4-phenyl-, (4R,4'R)-; (195433-00-2), **80**, 46

Dibenzoyl peroxide: Peroxide, dibenzoyl; (94-36-0), **80**, 75

Dibenzyl phosphate: Phosphoric acid, bis(phenylmethyl) ester; (1623-08-1), **80**, 219

Di-tert-butyl dicarbonate: Dicarbonic acid, bis(1,1-dimethylethyl) ester; (24424-99-5), **80**, 18

1,3-Dichloroacetone: 2-Propanone, 1,3-dichloro-; (534-07-6), **80**, 144

Dicobalt octacarbonyl: Cobalt, di-μ-carbonylhexacarbonyl di-, (Co-Co); (10210-68-1), **80**, 93

1,3-Dicyclohexylcarbodiimide: Cyclohexanamine, N,N'-methanetetraylbis-; (538-75-0), **80**, 219

(Z)-1,2-DIDEUTERIO-1-(TRIMETHYLSILYL)-1-HEXENE: Silane, (1,2-dideuterio-1-hexenyl)trimethyl-, (Z)-; **80**, 120

Diethylaminosulfur trifluoride (DAST): Sulfur, (N-ethylethanaminato)trifluoro-, (T-4); (38078-09-0), **80**, 46

Diethylaminotrimethylsilane: Silanamine, N,N-diethyl-1,1,1-trimethyl-; (996-50-9), **80**, 195

Diisobutylaluminium hydride: Aluminum, hydrobis(2-methylpropyl)-; (1191-15-7), **80**, 177

Diisopropylethylamine; 2-Propanamine, N-ethyl-N-(1-methylethyl)-; (7087-68-5), **80**, 207

1,2:4,5-Di-O-isopropylidene-β-D-fructopyranose: β-D-Fructopyranose, 1,2;4,5-bis-O-(1-methylethylidene)-; (25018-67-1), **80**, 1

1,2:4,5-DI-O-ISOPROPYLIDENE-D-erythro-2,3-HEXODIULO-2,6-PYRANOSE: β-D-erythro-2,3-Hexodiulo-2,6-pyranose, 1,2:4,5-bis-O-(1-methylethylidene)-; (18422-53-2), **80**, 1, 9

1,2-Dimethoxyethane: Ethane, 1,2-dimethoxy-; (110-71-4), **80**, 93

Dimethoxymethane: Methane, dimethoxy-; (109-87-5), **80**, 9

9,10-DIMETHOXYPHENANTHRENE: Phenanthrene, 9,10-dimethoxy-; (13935-65-4), **80**, 227

2,2-Dimethoxypropane: Propane, 2,2-dimethoxy-; (77-76-9), **80**, 1

Dimethylaluminum chloride: Aluminum, chlorodimethyl-; (1184-58-3), **80**, 133

4-Dimethylaminopyridine: 4-Pyridinamine, N,N-dimethyl-; (1122-58-3), **80**, 177

cis-5-(5,5-DIMETHYL-1,3-DIOXAN-2-YLIDENE)HEXAHYDRO-1(2H)-PENTALENONE, **80**, 144

N,N-Dimethylformamide: Formamide, N,N-dimethyl-; (68-12-2), **80**, 46, 200

6,6-DIMETHYL-1-METHYLENE-4,8-DIOXASPIRO[2.5]OCTANE: 4,8-Dioxaspiro [2.5]octane, 6,6-dimethyl-1-methylene; (122968-05-2), **80**, 144

3R,7-DIMETHYL-2-(2-OXOBUTYL)-6-OCTENAL: 6-Octenal, 3,7-dimethyl-2-(3-oxo-butyl)-, (3R)-; (131308-24-2), **80**, 195

Dimethyl sulfate: Sulfuric acid, dimethyl ester; (77-78-1), **80**, 227

DIPHENYL DISULFIDE: Disulfide, diphenyl; (882-33-7), **80**, 184

(1S,2R)-1,2-Diphenyl-2-hydroxyethylamine: Benzeneethanol, β-amino-α-phenyl-, (αR,βS)-rel-; (23412-95-5), **80**, 18

(1R,2S)-1,2-Diphenyl-2-hydroxyethylamine: Benzeneethanol, β-amino-α-phenyl-, (αS,βR)-; (23364-44-5), **80**, 18

N-(Diphenylmethylene)glycine tert-butyl ester: Glycine, (diphenylmethylene)-, 1,1-dimethylethyl ester; (81477-94-3), **80**, 38

Disodium ethylenediaminetetraacetate: Glycine, N,N'-1,2-ethanediylbis [N-(carboxymethyl)-, disodium salt; (139-33-3), **80**, 9

4-Dodecylbenzenesulfonyl azide: Benzenesulfonyl azide, 4-dodecyl-; (79791-38-1), **80**, 160

Ethyl bromoacetate: Acetic acid, bromo-, ethyl ester; (105-36-2), **80**, 18

Ethyl (1'R,2'S)-N-tert-butyloxycarbonyl-N-(1',2'-diphenyl-2'-hydroxyethyl) glycinate: Glycine, N-[(1,1-dimethylethoxy)carbonyl]-N-(2-hydroxy-1,2-diphenylethyl)-, ethyl ester, [S-(R*,S*)]-; (112741-73-8), **80**, 18

Ethyl (1'S,2'R)-N-tert-butyloxycarbonyl-N-(1',2'-diphenyl-2'-hydroxyethyl)glycinate: Glycine, N-[(1,1-dimethylethoxy)carbonyl]-N-(2-hydroxy-1,2-diphenylethyl)-, ethyl ester, [R-(R*,S*)]-; (112741-70-5), **80**, 18

Ethyl (1'S,2'R)-N-(1',2'-diphenyl-2'-hydroxyethyl)glycinate: Glycine, N-(2-hydroxy-1,2-diphenylethyl)-, ethyl ester, [R-(R*,S*)]-; (100678-82-8), **80**, 18

Ethyl (1'R,2'S)-N-(1',2'-diphenyl-2'-hydroxyethyl)glycinate: Glycine, N-(2-hydroxy-1,2-diphenylethyl)-, ethyl ester, [S-(R*,S*)]-; (112835-62-8), **80**, 18

Ethylenediaminetetraacetic acid (EDTA): Glycine, N,N'-1,2-ethanediylbis [N-(carboxymethyl)-;(60-00-4), **80**, 9

Ferric nitrate nonahydrate: Nitric acid, iron(3+) salt, nonahydrate; (7782-61-8), **80**, 144

2-Fluorosulfonyl-2,2-difluoroacetic acid: Acetic acid, difluoro(fluorosulfonyl)-; (1717-59-5), **80**, 172

D-Fructose; (57-48-7), **80**, 1

Furfural: 2-Furancarboxaldehyde; (98-01-1), **80**, 66

Hexacarbonyl[μ[(3,4-η:3,4-η)-2-methyl-3-butyn-2-ol]]dicobalt: Cobalt, hexacarbonyl [μ-[(3,4-η:3,4-η)-2-methyl-3-butyn-2-ol]]di-, (Co-Co); (40754-33-4), **80**, 93

Hexafluorophosphoric acid: Phosphate (1-), hexafluoro-, hydrogen; (16940-81-1), **80**, 200

1,1,1,3,3,3-Hexamethyldisilazane: Silanamine,1,1,1-trimethyl-N-(trimethylsilyl)-; (999-97-3), **80**, 160

Hydriodic acid; (10034-85-2), **80**, 129

Hydrogen peroxide (H_2O_2); (7722-84-1), **80**, 9, 184

(S)-[1-(Hydroxydiphenylmethyl)-2-methylpropyl]carbamic acid, tert-butyl ester: Carbamic acid, [1-(hydroxydiphenylmethyl)-2-methylpropyl]-, 1,1-dimethylethyl ester, (S)-; (157035-82-0), **80**, 57

2-Hydroxy-2(5H)-furanone: 2(5H)-Furanone, 5-hydroxy-; (14032-66-7), **80**, 66

Hydroxylamine hydrochloride: Hydroxylamine, hydrochloride; (5470-11-1), **80**, 207

N-HYDROXY-(S)-1-PHENYLETHYLAMINE OXALATE; Benzenemethanamine, N-hydroxy-α-methyl-, (αS)-, ethanedioate (1:1) salt; (78798-33-1), **80**, 207

1-HYDROXY-3-PHENYL-2-PROPANONE: 2-Propanone, 1-hydroxy-3-phenyl-; (4982-08-5), **80**, 190

(E)-3-IODOPROP-2-ENOIC ACID: 2-Propenoic acid, 3-iodo-, (2E)-; (6372-02-7), **80**, 129

Isoprene: 1,3-Butadiene, 2-methyl-; (78-79-5), **80**, 133

Isopropyl acetate: Acetic acid, 1-methylethyl ester; (108-21-4), **80**, 219

Isopropylmagnesium chloride: Magnesium, chloro(1-methylethyl)-; (1068-55-9), **80**, 120

Lithium; (7439-93-2), **80**, 31

Lithium bis(trimethylsilyl)amide; Lithium hexamethyldisilazide: Silanamine, 1,1,1-trimethyl-N-(trimethylsilyl)-, lithium salt; (4039-32-1), **80**, 31, 160

Magnesium; (7439-95-4), **80**, 57

Meldrum's acid: 1,3-Dioxane-4,6-dione, 2,2-dimethyl-; (2033-24-1), **80**, 133 d-Menthol: Cyclohexanol, 5-methyl-2-(1-methylethyl)-, (1S,2R,5S)-; (15356-60-2), **80**, 66

(5S)-(d-MENTHYLOXY)-2(5H)-FURANONE: 2(5H)-Furanone, 5-[[(1S,2R,5S)-5-methyl-2-(1-methylethyl)cyclohexyl]oxy]-, (5S)-; (122079-41-8), **80**, 66

trans-1-(4-METHOXYPHENYL)-4-PHENYL-3-PHENYLTHIO)AZETIDIN-2-ONE: 2-Azetidinone, 1-(4-methoxyphenyl)-4-phenyl-3-(phenylthio)-trans-; (94612-48-3), **80**, 160

Methyl acrylate: 2-Propenoic acid, methyl ester; (96-33-3), **80**, 38

Methyl 2-bromomethyl-3-nitrobenzoate: Benzoic acid, 2-bromomethyl-3-nitro-, methyl ester; (98475-07-1), **80**, 75

2-Methyl-3-butyn-2-ol: 3-Butyn-2-ol, 2-methyl-; (115-19-5), **80**, 93

Methyl 2-ethenyl-3-nitrobenzoate: Benzoic acid, 2-ethenyl-3-nitro-, methyl ester; (195992-04-2), **80**, 75

(4S)-(1-METHYLETHYL)-5,5-DIPHENYLOXAZOLIDIN-2-ONE: 2-Oxazolidinone, 4-(1-methylethyl)-5,5-diphenyl-, (4S)-; (184346-45-0), **80**, 57

Methyl 5-hexenoate: 5-Hexenoic acid, methyl ester; (2396-80-7), **80**, 111

Methyl iodide: Methane, iodo-; (74-88-4), **80**, 57, 144

METHYL INDOLE-4-CARBOXYLATE: 1H-Indole-4-carboxylic acid, methyl ester; (39830-66-5), **80**, 75

Methyl 2-methyl-3-nitrobenzoate: Benzoic acid, 2-methyl-3-nitro-, methyl ester; (59382-59-1), **80**, 75

METHYL 5-METHYLPYRIDINE-2-CARBOXYLATE: 2-Pyridinecarboxylic acid, 5-methyl-, methyl ester; (260998-85-4), **80**, 133

METHYL PHENYL SULFOXIDE: Benzene, (methylsulfinyl)-; (1193-82-4), **80**, 184

trans-β-Methylstyrene: Benzene, (1E)-1-propenyl-; (873-66-5), **80**, 9

(R,R)-trans-β-METHYLSTYRENE OXIDE: Oxirane, 2-methyl-3-phenyl-, (2R,3R)-; (14212-54-5), **80**, 9

Methyl vinyl ketone: 3-Buten-2-one; (78-94-4), **80**, 195

Neopentyl glycol: 1,3-Propanediol, 2,2-dimethyl-; (126-30-7), **80**, 144

Oxalic acid: Ethanedioic acid; (144-62-7), **80**, 207

Oxone: Peroxymonosulfuric acid, monopotassium salt, mixture with dipotassium sulfate and potassium hydrogen sulfate; (37222-66-5), **80**, 9

Oxygen; (7782-44-7), **80**, 66

Palladium acetate: Acetic acid, palladium(2+) salt; (3375-31-3), **80**, 75

Paraformaldehyde; (30525-89-4), **80**, 75

Perchloric acid; (7601-90-3), **80**, 1

9,10-Phenanthrenequinone; (84-11-7), **80**, 227

1-Phenylcyclohexene: Benzene, 1-cyclohexen-1-yl-; (771-98-2), **80**, 9

(R,R)-1-PHENYLCYCLOHEXENE OXIDE: 7-Oxabicyclo[4.1.0]heptane, 1-phenyl-, (1R,6R)-; (17540-04-4), **80**, 9

S-Phenyl diazothioacetate: Ethanethioic acid, diazo-, S-phenyl ester; (72228-26-3), **80**, 160

(S)-1-Phenylethylamine: Benzenemethanamine, α-methyl-, (αS)-; (2627-86-3), **80**, 207

(S)-[(1-Phenylethyl)amino]acetonitrile; Acetonitrile, [[(1S)-1-phenylethyl]amino]-; (35341-76-5), **80**, 207

[(1S)-1-Phenylethyl]imino]acetonitrile N-oxide; Acetonitrile, [oxido[(1S)-1-phenylethyl]imino]-; (300843-73-6), **80**, 207

(R)-(–)-2-Phenylglycinol: Benzeneethanol, β-amino-, (βR)-; (56613-80-0), **80**, 46

Phenylmagnesium bromide: Magnesium, bromophenyl-; (100-58-3), **80**, 57

3-Phenyl-2-propylthio-2-propen-1-ol, **80**, 190

3-Phenyl-2-propyn-1-ol: 2-Propyn-1-ol, 3-phenyl-; (1504-58-1), **80**, 190

S-Phenyl thioacetate: Ethanethioic acid, S-phenyl ester; (934-87-2), **80**, 160

Phosphorus oxychloride: Phophoric trichloride; (10025-87-3), **80**, 200

Tetrakis(hydroxymethyl)phosphonium sulfate ("Pyroset-TKOW"): Phosphonium, tetrakis(hydroxymethyl)-, sulfate (2:1); (55566-30-8), **80**, 85

1-Tetralone: 1(2H)-Naphthalenone, 3,4-dihydro-; (529-34-0), **80**, 104

N,N,N',N'-Tetramethylethylenediamine: 1,2-Ethanediamine, N,N,N',N'-tetramethyl-; (110-18-9), **80**, 46

Thioanisole: Benzene, (methylthio)-; (100-68-5), **80**, 184

Thionyl chloride; (7719-09-7), **80**, 46

Titanium tetraisopropoxide: Titanium, tetrakis(2-propanolato)-, 2-Propanol, titanium (4+); (546-68-9), **80**, 120

p-Toluenesulfonic acid: Benzenesulfonic acid, 4-methyl-; (104-15-4), **80**, 18, 144

p-Toluenesulfonyl chloride: Benzenesulfonyl chloride, 4-methyl-; (98-59-9), **80**, 93, 133

5-(Tosyloxyimino)-2,2-dimethyl-1,3-dioxane-4,6-dione: 1,3-Dioxane-4,5,6-trione, 2,2-dimethyl-, 5-O-[(4-methylphenyl)sulfonyl]oxime; (215436-24-1), **80**, 133

8-[2-(TRIETHOXYSILYL)ETHYL-1-TETRALONE: 1(2H)-Naphthalenone, 3,4-dihydro-8-[2-(triethoxysilyl)ethyl]-; (154735-94-1), **80**, 104

Triethoxyvinylsilane: Silane, ethenyltriethoxy-; (78-08-0), **80**, 104

Triethylamine: Ethanamine, N,N-diethyl-; (121-44-8), **80**, 18, 46, 75, 85, 161, 233

Triethylsilane: Silane, triethyl-; (617-86-7), **80**, 93

Trifluoroethanol: Ethanol, 2,2,2-trifluoro-; (75-89-8), **80**, 184

2,2,2-Trifluoroethyl trifluoroacetate: Acetic acid, trifluoro-, 2,2,2-trifluoroethyl ester; (407-38-5), **80**, 160

Triiospropyl triflate: Methanesulfonic acid, trifluoro-, tris(1-methylethyl)silyl ester; (80522-42-5), **80**, 233

1,6,6-Trimethyl-4,8-dioxaspiro[2.5]oct-1-ene: 4,8-Dioxaspiro[2.5]oct-1-ene, 1,6,6-trimethyl-; (122762-81-6), **80**, 144

1-(Trimethylsilyl)-1-hexyne: Silane, 1-hexynyltrimethyl-; (3844-94-8), **80**, 120

Triphenylphosphine: Phosphine, triphenyl-; (603-35-0), **80**, 75

Printed and bound by CPI Group (UK) Ltd, Croydon, CR0 4YY

16/04/2025

14658347-0002